Springer Theses

Recognizing Outstanding Ph.D. Research

Aims and Scope

The series "Springer Theses" brings together a selection of the very best Ph.D. theses from around the world and across the physical sciences. Nominated and endorsed by two recognized specialists, each published volume has been selected for its scientific excellence and the high impact of its contents for the pertinent field of research. For greater accessibility to non-specialists, the published versions include an extended introduction, as well as a foreword by the student's supervisor explaining the special relevance of the work for the field. As a whole, the series will provide a valuable resource both for newcomers to the research fields described, and for other scientists seeking detailed background information on special questions. Finally, it provides an accredited documentation of the valuable contributions made by today's younger generation of scientists.

Theses are accepted into the series by invited nomination only and must fulfill all of the following criteria

- They must be written in good English.
- The topic should fall within the confines of Chemistry, Physics, Earth Sciences, Engineering and related interdisciplinary fields such as Materials, Nanoscience, Chemical Engineering, Complex Systems and Biophysics.
- The work reported in the thesis must represent a significant scientific advance.
- If the thesis includes previously published material, permission to reproduce this must be gained from the respective copyright holder.
- They must have been examined and passed during the 12 months prior to nomination.
- Each thesis should include a foreword by the supervisor outlining the significance of its content.
- The theses should have a clearly defined structure including an introduction accessible to scientists not expert in that particular field.

More information about this series at http://www.springer.com/series/8790

Mickey McDonald

High Precision Optical Spectroscopy and Quantum State Selected Photodissociation of Ultracold 88Sr2 Molecules in an Optical Lattice

Doctoral Thesis accepted by Columbia University, New York, USA

Mickey McDonald
University of Chicago
James Franck Institute
Chicago, IL, USA

ISSN 2190-5053 ISSN 2190-5061 (electronic)
Springer Theses
ISBN 978-3-319-68734-6 ISBN 978-3-319-68735-3 (eBook)
https://doi.org/10.1007/978-3-319-68735-3

Library of Congress Control Number: 2017953964

© Springer International Publishing AG 2018

This work is subject to copyright. All rights are reserved by the Publisher, whether the whole or part of the material is concerned, specifically the rights of translation, reprinting, reuse of illustrations, recitation, broadcasting, reproduction on microfilms or in any other physical way, and transmission or information storage and retrieval, electronic adaptation, computer software, or by similar or dissimilar methodology now known or hereafter developed.

The use of general descriptive names, registered names, trademarks, service marks, etc. in this publication does not imply, even in the absence of a specific statement, that such names are exempt from the relevant protective laws and regulations and therefore free for general use.

The publisher, the authors and the editors are safe to assume that the advice and information in this book are believed to be true and accurate at the date of publication. Neither the publisher nor the authors or the editors give a warranty, express or implied, with respect to the material contained herein or for any errors or omissions that may have been made. The publisher remains neutral with regard to jurisdictional claims in published maps and institutional affiliations.

Printed on acid-free paper

This Springer imprint is published by Springer Nature
The registered company is Springer International Publishing AG
The registered company address is: Gewerbestrasse 11, 6330 Cham, Switzerland

To Bart McGuyer, who taught me what it means to be a scientist.

Supervisor's Foreword

This is an exciting time for physics, and particularly for the field of atomic and molecular physics. Scientists can exert nearly perfect control over atomic motion and quantum states and use this exquisite precision to seek answers to fundamental questions. In recent years, there has been a surge in growth of cold and ultracold molecular science. Molecules are in many ways similar to atoms but offer greatly expanded possibilities for understanding a wide array of physical phenomena. This, of course, comes with new challenges, since even diatomic molecules have much more complicated spectra than atoms.

Mickey McDonald's thesis work unites studies of basic molecular quantum physics with the high-precision techniques of ultracold-atom optical lattice clocks. His sophisticated experiment involves microkelvin diatomic strontium molecules in an optical lattice trap and utilizes optical spectroscopy to study physics near the atom-molecule threshold. Mickey obtained unique lattice clock-style spectroscopic measurements that clearly demonstrate the potential of ultracold molecule science and open the door to table-top tests of fundamental physics, such as molecular quantum electrodynamics and nanometer-scale deviations from Newtonian gravity, at a level that was not possible before.

Mickey's thesis carefully demonstrates how forbidden transitions in molecules can be enabled with much weaker magnetic fields than in their constituent atoms, with implications for metrology and precision measurement. It also describes work on two-body subradiance, reporting by far the deepest subradiance to date, a 300-fold suppression of radiative emission. This is a fundamental two-particle quantum optics effect, and it turns out that homonuclear diatomic molecules in an optical lattice are the ideal system for studying its properties.

Furthermore, Mickey's thesis describes pioneering work on ultracold chemistry via studies of molecular photodissociation. At extremely low temperatures, processes that create or break chemical bonds proceed according to quantum mechanical rules, and several approaches had been tried to observe these non-classical phenomena. In our lab, we had been detecting weakly bound molecules by fragmenting them into atoms, which were then imaged with a camera facing perpendicular to the lattice trapping axis. Mickey was curious what would happen if we aligned a camera on axis with the lattice and looked at the photofragment angular distributions. So he led an effort to build this setup and learn how to process and interpret the images. The textbook-quality images shown in this thesis are beautiful pictures of diatomic molecules breaking apart in up to eight different directions, exhibiting coherent quantum mechanical patterns. This nonclassical behavior challenged intuition that was established before the 1980s. The thesis demonstrates how to break molecular bonds while imparting only a miniscule level of excess energy to the fragments and thoroughly explores the cold regime from 100 nK to tens of millikelvin. The results show matter-wave interference of reaction products as well as quantum mechanical reaction barrier tunneling.

Measuring ultralow temperatures, down to the nanokelvin level, is in itself an important outstanding problem, and Mickey's thesis describes an invention of an excellent technique that relies purely on spectroscopy, thus yielding very high precision. It only requires the atoms or molecules to be tightly trapped in a lattice and to possess a narrow transition. Mickey uses this technique to characterize the temperature of ultracold strontium molecules and even to cool them down with a new technique of "carrier cooling."

The pioneering experiments described in this thesis set the stage for optical precision measurements with simple molecules at a level of quantum control that was previously impossible. With this new toolkit of controlling and measuring molecular properties at an unprecedented level, we can be sure to unlock many additional mysteries of molecular and fundamental physics.

New York, NY, USA   Tanya Zelevinsky
July 2017

Acknowledgments

Writing a thesis can be a difficult, despair-ridden thing, marked by lonely nights and waves of self-doubt. And in the midst of this undertaking, it's easy to lose perspective. In some ways, the thesis itself seems silly and ill-defined. Is it really possible to distill 6 years of life experience into 200 pages? Is that even the most efficient way to transfer knowledge? And who in their right mind would actually *read* such a thing? Should the thesis be a survey of the field? An original piece of unpublished research? A technical manual for the next generation of graduate students? An opportunity to wax poetic on the meaning of life and how it all relates to ultracold molecules?

The fact is that no one tells you what the thesis is supposed to be. You just write it, and hope that someone *somewhere* will find it useful. Such an open-ended task can be daunting, and I've found myself at times overwhelmed by the immensity of what lay before me. But as I write these acknowledgements, I'm sitting on a plane bound from Korea to New York, just getting back from presenting work at my eighth international conference. I literally get paid to think about how to solve the mysteries of the universe and interact daily with some of the smartest people on the planet. Objectively, I am incredibly fortunate. A physicist's job is difficult, but at the same time it feels like a stretch to call the profession "work." My 6 years at Columbia have been marked by a constant pursuit of answers to questions about how the universe behaves. I'm grateful to have been given the opportunity to engage in this kind of work, because I know how rare it is to be able to get paid to do what you love.

When I first came to Columbia in the fall of 2010, I had the unfortunate combination of a big ego and zero skills. At my first meeting with Tanya Zelevinsky's group, Tanya introduced me and asked for Gael Reinaudi and Chris Osborn (the group's lead postdoc and first grad student respectively) to try and find something for me to do. After asking whether I had ever built a circuit ("No..."), possessed any programming experience ("Matlab 101?"), or knew anything about lasers ("...at Cornell I studied particle physics?"), Gael suggested "Well, perhaps you could rest your hand on this stack of papers and make sure they don't blow away in the wind."

Despite this ominous beginning, Gael and Chris eventually took me under their wings and showed me how to build lasers, run the experiment, and troubleshoot thorny problems. My success at Columbia wouldn't have been possible without Gael's enthusiasm and Chris's patience. I was lucky to inherit an experiment already largely built by these two great scientists over the previous 4 years. Without this head start, our results over the next few years wouldn't have been nearly as dramatic.

Near the end of Chris and Gael's time at ZLab, Bart McGuyer came to Columbia to take over leadership of the strontium experiment. After Chris left I became the senior grad student working on strontium, meaning that Bart and I would be collaborating closely. And whereas I had a fragile ego and was driven largely by passion and obsession in choosing what questions to pursue, Bart was levelheaded and careful, devoted to getting the details right. He was also (much to my dismay) clearly quite a bit smarter than me.

I remember once early in my time at Columbia getting into some physics-related argument with Bart and having him clearly and calmly (but also quite bluntly) telling me exactly why I was wrong about some finer detail of the analysis of a recent experiment. My fragile ego was once again sorely bruised, and after the conversation I basically stormed out to clear my head with a coffee and a breath of fresh air. And I remember talking to Chris later about this, and complaining about how Bart was totally unwilling to phrase his criticisms more delicately. To this he replied: "Mickey, I don't think you realize how lucky you are to have Bart here as your postdoc. He's a *real* scientist." It took me another year to fully appreciate this fact, but I now see just how prescient Chris's words were. This thesis wouldn't have been possible without Bart, nor would I be half the scientist I am today without having had him here pushing me to be better. As I move on to my own postdoc, I have Bart's example to follow, and can only hope that I can offer the same kind of mentorship to my own future graduate students as Bart did for me. And I hope Bart won't be too embarrassed when he sees I've dedicated this thesis to him.

I've had the great pleasure of working with a series of outstanding high school, undergraduate, and master's students during my time at Columbia, who indulged my (sometimes poorly thought out) curiosities and inspired me to think more deeply and to try harder to understand what I didn't yet know so well. High school senior Jennifer Ha helped me finally realize an old idea for a lecture demonstration of the wave nature of light, leading to a paper we published in the *American Journal of Physics* [1]. Noel Wan and Elijahu Ben-Michael spent countless hours fiddling with prototypes for a crazy idea of mine for a "Laser-light intensity-insensitive passive polarization-stabilization setup (LLIIPPSS)," which was unfortunately doomed by subtly persistent interference effects (but thanks for the great work guys!). Matthew Miecnikowski helped realize another idea for an optics experiment with pedagogical value: if frequency shifts in acousto-optic modulators are due to first-order Doppler shifts from a diffraction grating moving at the speed of sound, is it possible to observe second-order shifts? (Once again, it turned out that unfortunately the answer was "No"... but we learned a lot!)

Acknowledgments

xi

I'm especially grateful to Florian Apfelbeck for his hard work and inspiring attitude as we worked together to study the angular distribution of fragments produced by the photodissociation of ultracold molecules, work which was recently published in *Nature* and is described in the last chapter of this thesis. This was a project which was really high risk, with no precedent in lattice-based precision spectroscopy experiments and only a tenuous connection to the molecular clock-based research goals of ZLab, but Florian embraced the challenge wholeheartedly and with great enthusiasm. The late nights we spent troubleshooting our optics and taking pictures of molecular explosions are some of the highlights of my PhD, and the master's thesis Florian wrote at the end of his stay with us still makes me proud whenever I open it.

When I leave Columbia, Geoff Iwata will inherit the title of "Senior Zlab Grad Student," I'm really grateful for his help during his first years in ZLab on the strontium experiment (several entries in the "Zeeman shifts" Tables are directly due to him!), and am consistently impressed with his leadership in building an entirely new experiment to explore the direct cooling of barium hydride molecules. Geoff's wonderful attitude made a dark, windowless lab a lot more inviting, and I credit his presence here to keeping me a little more sane. And by the way, thank you for loaning me your computer when mine died 1 week before my defense!

Chih-Hsi Lee came to Columbia in 2014 and has already taken over the strontium experiment. This period of transition, with both the lead graduate student and senior postdoc leaving at about the same time, was potentially a recipe for disaster, with a strong possibility that progress on the experiment could halt with the loss of a lot of arcane knowledge. But Chih-Hsi turned out to be exactly the right person to pick up the reigns and forge ahead. I remember that after late nights of data taking, Florian and I would sometimes go for a drink, and the conversation on *multiple* occasions drifted toward how smart Chih-Hsi was and how lucky we were to have him. I'm especially grateful for his ability to put up with me and all my moodiness during this final year of thesis writing and job searching. I'm sure he'll take the experiment in directions Bart and I wouldn't have anticipated, and I just can't wait to hear about all the discoveries he'll make in the future.

Many of our most interesting results wouldn't have been possible without the collaboration of Robert Moszynski and his students Wojtek Skomorowski and Iwona Majewska. Once again I find myself marvelling at my good fortune for having joined an experiment with such a close collaboration with such great theorists. Our most precise measurements are only interesting if they can be placed in some sort of theoretical context, whether that context is precise comparison with ab initio calculations or more general rules derived from stripped-down models. Robert's group was instrumental in providing the brainpower to fuel both of these kinds of understanding. Thanks to both Wojtek and Robert for their willingness to answer my "simple questions" regarding the structure of strontium dimers, and to Iwona for her quick work in understanding the mathematics of photofragment angular distributions.

There really are too many people to thank for making ZLab a great place to work, and though I have stories about everyone, I don't think I can devote a dozen pages here to their telling. But let me thank specifically Rees McNally, Fabian Soerensen, Cole Rainey, Damon Daw, Leila Abbih, and Katerina Verteletsky for helping make this lab a happy place to work. And thank you Marco Tarallo for helping make sure it wasn't *too* happy. (I'm pretty sure Marco will laugh at that one...I hope...) Thanks also to Marco for indulging me by listening to my skepticism about LIGO and crazy ideas about the expansion of the universe. And also...go Bills?

Thanks to Dr. Andrew Grier for his hard work in helping set up our frequency comb and his genuinely caring attitude. Andrew is a great scientist, and I think Chih-Hsi will find his life made a lot easier due to the foundation laid by Andrew.

Of course, none of these great students and scientists would have met one another had it not been for the lab which Tanya Zelevinsky has set up. I'm very grateful for having landed here, and especially for Tanya's support of my crazy ideas. I've spent a lot of time at Columbia pursuing ideas which ultimately did not produce fruitful results, as well as a minority of time pursuing ideas which did. I know I'm very lucky to have had an advisor who was willing to let me go off in crazy directions and see where they might take me. This policy is what led to the results of our recent *Nature* paper and is one I'll remember as I go off to lead my own research projects. So thank you Tanya for leading this great lab and allowing me to be a part of it.

Finally, I'd like to thank my family for providing a constant source of love and support during a very stressful 6 years. I decided to come to Columbia in large part because of the possibility of being close to my brother and sister, and I know now that this decision was the right one. From Bagel Sundays to dinner at Connor and Katie's to Christmases back in Fairport, we've had some pretty great times. You've provided a constant, secure source of support, and knowing that I could rely on that made it a whole lot easier to throw myself completely into my research. I make no exaggeration when I say that without Erin, Connor, Mom, and Dad, this thesis wouldn't have been possible.

Reference

1. McDonald, M., Ha, J., McGuyer, B.H., Zelevinsky, T.: Visible optical beats at the hertz level. Am. J. Phys. **82**(10), 1003–1005 (2014)

Parts of this thesis have been published in the following journal articles:

1. McDonald, M., McGuyer, B. H., Apfelbeck, F., Lee, C. H., Majewska, I., Moszynski, R., & Zelevinsky, T. Photodissociation of ultracold diatomic strontium molecules with quantum state control. *Nature*, **535**, 7610 (2016), 122–133.
2. McGuyer, B. H., McDonald, M., Iwata, G. Z., Tarallo, M. G., Grier, A. T., Apfelbeck, F., & Zelevinsky, T. High-precision spectroscopy of ultracold molecules in an optical lattice. *New Journal of Physics*, **17**, 5 (2015), 055004.
3. McDonald, M., McGuyer, B. H., Iwata, G. Z., & Zelevinsky, T. Thermometry via light shifts in optical lattices. *Physical Review Letters*, **114**, 2 (2015), 023001.
4. McGuyer, B. H., McDonald, M., Iwata, G. Z., Skomorowski, W., Moszynski, R., & Zelevinsky, T. Control of optical transitions with magnetic fields in weakly bound molecules. *Physical Review Letters*, **115**, 5 (2015), 053001.
5. McGuyer, B. H., McDonald, M., Iwata, G. Z., Tarallo, M. G., Skomorowski, W., Moszynski, R., & Zelevinsky, T. Precise study of asymptotic physics with subradiant ultracold molecules. *Nature Physics*, **11**, 1 (2015), 32–36.
6. McGuyer, B. H., Osborn, C. B., McDonald, M., Reinaudi, G., Skomorowski, W., Moszynski, R., & Zelevinsky, T. Nonadiabatic effects in ultracold molecules via anomalous linear and quadratic Zeeman shifts. *Physical Review Letters*, **111**, 24 (2013), 243003.
7. Reinaudi, G., Osborn, C. B., McDonald, M., Kotochigova, S., & Zelevinsky, T. Optical Production of Stable Ultracold $^{88}Sr_2$ Molecules. *Physical Review Letters*, **109**, 11 (2012), 115303.

Contents

1 **Introduction** 1
   1.1 Why Build a Molecular Clock? And How to Go About It? 2
       1.1.1 A New Kind of Clock 3
   1.2 A Testbed for Quantum Chemistry 3
   1.3 This Thesis 4
   References 6

2 **Structure of ^{88}Sr and ^{88}Sr$_2$** 9
   2.1 Rotational Structure 10
       2.1.1 Hund's Cases 11
       2.1.2 Electronic Ground State 12
       2.1.3 Electronic Excited State 13
   References 15

3 **Measurements of Binding Energies** 17
   3.1 Catalog of All Known Rovibrational Levels 18
       3.1.1 $^1S_0+^3P_1$ 19
       3.1.2 $^1S_0+^1S_0$ 19
   3.2 Photo*association* (Free-Bound) Spectroscopy 21
       3.2.1 Observing Photoassociation Losses 21
       3.2.2 One-Photon (Electronic Excited State) 24
       3.2.3 Two-Photon (Electronic Ground State) 26
   3.3 Photo*dissociation* (Bound-Free) Spectroscopy 28
       3.3.1 Observing Photodissociation Gains 31
       3.3.2 Recovery Transitions 32
       3.3.3 Direct One-Photon Photodissociation 33
       3.3.4 Direct Two-Photon Photodissociation 38
   3.4 Bound-Bound Spectroscopy 45
       3.4.1 Case Study: $1_g(-v, J'=1)$ 45

		3.5 Coherent Two-Photon Raman Transitions (Ground State Binding Energy Differences)	51
		3.5.1 Comments on Coherence Time	51
		3.5.2 Determination of Binding Energy Differences Among $X(-1, 0)$, $X(-2, 0)$, and $X(-3, 0)$ States	53
	References		56
4	**Measurements of Zeeman Shifts**		59
	4.1	Introduction and Summary of Measurements	59
		4.1.1 Details About Data Presented in Tables 4.1, 4.2, 4.3, and 4.4...	60
	4.2	Linear Zeeman Shifts (Low-Field)	62
		4.2.1 Calculation of the Linear Zeeman Shift of the $^3P_1$ State of ^{88}Sr	63
		4.2.2 'Ideal' Linear Zeeman Shifts for Molecules Satisfying Hund's Case (c)	64
		4.2.3 Validity of Hund's Case (c).....................................	71
	4.3	Quadratic (and Higher Order) Zeeman Shifts	71
		4.3.1 Option 1: Coriolis Coupling of the $\Omega = 0, 1$ Potentials	71
		4.3.2 Option 2: Decreased Level Spacing Near the Top of the Potential	73
		4.3.3 Comparison of Atomic and Molecular Quadratic Zeeman Shifts..	75
		4.3.4 Determination of Higher (up to Sixth) Order Zeeman Shifts	79
	4.4	Description of Magnetic Field Coils	79
		4.4.1 Calibration to the $^1S_0 + ^3P_1$ Intercombination Line	82
		4.4.2 Quantized Output from NI PXI-6713 Card....................	83
	References		84
5	**Magnetic Control of Transition Strengths**		87
	5.1	Introduction (Defining "Transition Strength")........................	87
	5.2	Three Ways to Measure	88
		5.2.1 Normalized Area Under a Lorentzian	88
		5.2.2 Rabi Oscillations...	93
		5.2.3 Autler-Townes Splitting (Two-Photon Spectroscopy)	96
	5.3	Enabling "Forbidden" Transitions with Magnetic Fields	98
		5.3.1 Intuitive Model Based on Perturbation Theory................	99
		5.3.2 Results...	101
	References		104
6	**Subradiant Spectroscopy**...		107
	6.1	Introduction: Subradiance vs Superradiance...........................	107
	6.2	Characterizing Transition Strengths....................................	108
		6.2.1 Aside: Isolating E1, M1, and E2 Transitions from One Another..	109

		6.2.2	Normalized Area Under a Lorentzian	110
		6.2.3	Rabi Oscillations	110
	6.3	Characterizing Linewidths (I): Sources of *Artificial* Broadening		114
		6.3.1	(I) Spectroscopy	114
		6.3.2	(II) "In the Dark" Lifetime Measurements	120
	6.4	Characterizing Linewidths (II): Sources of *Natural* (i.e., Inherent) Broadening		122
		6.4.1	Radiative Decay	123
		6.4.2	Predissociation	126
		6.4.3	Results: Linewidth vs Bond Length at Zero Field	129
		6.4.4	Magnetic Field Mixing of Nearby Levels	131
	6.5	Comment on the Search for 0_g States		132
	References			133
7	**Carrier Thermometry in Optical Lattices**			135
	7.1	A Roadmap for Determining Temperature		136
	7.2	Overview: Setting Up the Problem		138
		7.2.1	Modeling the Trap Potential	139
		7.2.2	Calculating Lineshapes	142
	7.3	Experimental Techniques and Results		152
		7.3.1	Recording High S/N Lineshapes	153
		7.3.2	Identifying Sources of Molecular Heating	157
	References			159
8	**Photodissociation and Ultracold Chemistry**			161
	8.1	A ZLab History of Photodissociation Measurements		161
	8.2	A Slightly Broader History of Photodissociation Measurements		165
		8.2.1	Early (Classical) Theory	165
		8.2.2	The First Experiments	165
		8.2.3	Difficulties in Comparing Experiment to a Fully Quantum Mechanical Theory	166
	8.3	Photodissociation of Ultracold $^{88}Sr_2$ Molecules in an Optical Lattice		167
		8.3.1	Imaging Photofragment Angular Distributions (PADs)	167
		8.3.2	Extracting Quantitative Angular Information	169
		8.3.3	Kinematics of a Photodissociation Reaction	169
		8.3.4	A Fully Quantum Mechanical Understanding of Photodissociation	175
	8.4	Preliminary Data and Unresolved Mysteries		176
		8.4.1	Magnetic Field Dependence of Dissociation Above the $1_u/0_u$ Barrier	176
		8.4.2	Frequency Dependence of One-Photon Dissociation of $J=1$ States Down to $^1S_0 + ^1S_0$ Threshold in the Absence of Shape Resonances	180
	References			181
Vitae				183

Chapter 1
Introduction

Over the past several decades, rapid progress has been made toward the accurate characterization and control of atoms, made possible largely by the development of narrow-linewidth lasers and techniques for trapping and cooling at ultracold temperatures. Extending this progress to molecules will have exciting implications for chemistry, condensed matter physics, and precision tests of physics beyond the Standard Model. These possibilities are all consequences of the richness of molecular structure, which is governed by physics substantially different from that characterizing atomic structure. This same richness of structure, however, increases the complexity of any molecular experiment manyfold over its atomic counterpart, magnifying the difficulty of everything from trapping and cooling to the comparison of theory with experiment.

This thesis describes work performed over the past 6 years to establish the state of the art in manipulation and quantum control of ultracold molecules. The molecules I discuss are very weakly bound (and therefore very large) ^{88}Sr$_2$ dimers, produced via photoassociation of ultracold strontium atoms followed by spontaneous decay to a stable ground state. We study their rovibrational structure from several different perspectives, including determinations of binding energies; linear, quadratic, and higher order Zeeman shifts; transition strengths between bound states; and lifetimes of narrow subradiant states. The physical intuition gained in these experiments applies generally to weakly bound diatomic molecules, and suggests extensive applications in precision measurement and metrology. In addition, I present a detailed analysis of the thermally broadened spectroscopic lineshape of molecules in a non-magic optical lattice trap, showing how such lineshapes can be used to directly determine the temperature of atoms or molecules in situ, addressing a long-standing problem in ultracold physics. Finally, I discuss the measurement of photofragment angular distributions produced by photodissociation, leading to an exploration of quantum-state-resolved ultracold chemistry.

1.1 Why Build a Molecular Clock? And How to Go About It?

Our work in the Zelevinsky lab straddles the border between two formerly unrelated fields. The first field is metrology, or precision measurements, a focus which is built into our lab's very DNA. Construction of our experiment first began in 2008 with the intention of adapting techniques originally intended for the development of optical atomic lattice clocks (which currently represent the best timekeepers in the world, so accurate that they can sense the gravitational potential difference induced by a height difference of 1 cm [9]), to the development of a new kind of clock entirely: a molecular lattice clock. Such a clock would use as its resonator not the oscillations of an electron between energy levels in an atom, but rather the relative vibrations of the two nuclei in a diatomic molecule. The second field is chemistry, but in the physicist's sense: stripping down a chemical reaction to its barest essentials, controlling every quantum mechanical degree of freedom, and using ab initio calculations to try and predict the behavior of reactions and the structure of molecules.

Why is this interesting? If atomic clocks are already so amazingly precise, is it really necessary to build a substitute? It turns out that this is a rather deep question. A clock is a device which measures time, ideally in a way which is independent of the local environment. But whether or not the clock's oscillator is decoupled from its environment is a tricky question to answer, and in fact requires comparison with another clock which is certain not to depend on the environmental factor in question.

To illustrate this, consider the pendulum clock, whose oscillator consists of a hanging mass swinging with a period T given by

$$T \approx 2\pi \sqrt{\frac{L}{g}}, \tag{1.1}$$

where L is the pendulum length and g is the local acceleration due to gravity. Immediately we can see that the period of a pendulum clock is in fact strongly coupled to the environment in that its period is inversely proportional to the square root of the local gravitational acceleration. But how would a scientist living before the discovery of Newton's Laws know this? If he were to bring such a pendulum clock to perform an experiment on Mt. Everest, would he realize that the times he measured were just a bit too short?

One way to discover this effect would be to bring two clocks: a pendulum, and another whose mechanism does not depend on the local gravity. Then, as the two clocks are brought to different heights, any relative drift between the two would be a sign of a change in the physics governing one of their oscillators. In this way, we've now turned the task of building clocks which depend on different physics into a technique for discovering *temporal or spatial variations in the laws of physics*.

This gets to the core of our desire to construct a molecular clock. The current record for accuracy in atomic clocks is impressive, but currently these claims of accuracy can only be checked by beating against other atomic clocks. Such comparisons between clocks that rely on similar resonator mechanisms could be blind to spatial or temporal changes in the physics governing those resonators. And while physicists work hard to minimize the influence of environmental effects, we do not know whether physics itself, embodied in natural constants parameterizing the strengths of interactions between nuclei and electrons, might be changing.

1.1.1 A New Kind of Clock

The primary physics governing the "tick rate" of an atomic clock is the strength of the attraction of an electron to the nucleus, embodied in the fine structure constant α. For a molecular clock, however, it turns out that the situation is different. Here, the clock's tick rate is most sensitive to the ratio of the masses of the nuclei to their electron clouds, embodied in the electron-to-proton mass ratio μ [2, 10]. Therefore building an extremely precise molecular clock serves also as a relatively model-independent method for determining whether or not the mass of the electron is drifting with respect to the mass of the proton.

Much of my work has focused on developing techniques to be eventually applied toward the construction of a molecular clock. In this thesis I'll describe work to understand lineshapes, a new technique for measuring temperature in optical lattices, precise determinations of the binding energies of weakly bound levels as well as their differences, and the discovery of super-narrow transitions to subradiant states, as well as a slew of other techniques and tricks which represent the state of the art in coherent control and interrogation of ultracold molecules.

1.2 A Testbed for Quantum Chemistry

While progress towards a molecular clock has always been our long-term goal, our day-to-day investigations into the properties of strontium molecules often stray into the regime of quantum chemistry and molecular physics. Indeed, our most fruitful collaboration to date has been with Robert Moszynski at the University of Warsaw in Poland, whose group has used our measurements to refine quantum chemistry models of the structure of $^{88}Sr_2$.

From an experimental point of view, making precise measurements which fall neatly on a theory curve can be extremely satisfying. But perhaps even more satisfying is discovering a new kind of quantity to measure. A molecule is more than the sum of its rovibrational levels, which is to say: there are many different experimental observables which can be used to characterize a molecule, besides simply its spectrum. The history of publications produced by ZLab is in part a history of discovering new quantities to measure, each of which tells part of a larger story of the structure of a molecule. In this thesis I'll describe precise measurements of molecular binding energies, Zeeman shifts, transition strengths, lifetimes, lineshapes, and photofragment angular distributions (PADs). While the precise values obtained from these measurements are interesting when they can be compared against theory, their dependence upon other molecular parameters, such as bond length, level spacing, or quantum numbers, allows for an intuition to be built up that deepens our general understanding of weakly bound two-body systems.

1.3 This Thesis

This thesis is divided into chapters which each focus on the precise measurement and physical interpretation of a different experimental observable. The measurements and techniques described herein summarize and expand upon work performed and published over the past 6 years [3–8]. A brief overview of what's covered in the following chapters is given below.

Structure of ^{88}Sr and ^{88}Sr$_2$ (Chap. 2) In many ways, ^{88}Sr$_2$ is the simplest molecule one could hope for. It consists of only two atoms, is homonuclear, and possesses zero nuclear spin. The advantage of working with a molecule so simple is that attempts to fully understand its structure from first principles become tenable. This chapter summarizes the structure of ^{88}Sr$_2$ in terms of Hund's cases, and defines the labeling scheme we will use to refer to different rovibrational levels. Knowledge of how quantum statistics influences the allowed values of certain quantum numbers will be important for interpreting several experiments described in later chapters.

Measurements of Binding Energies (Chap. 3) We have observed molecular resonances corresponding to every rovibrational level in ^{88}Sr$_2$ with a binding energy less than 8.5 GHz and rotational angular momentum $J \leq 4$ (except for those occupying the 0_g potential, which remain to be observed). Nearly every measurement is made with sub-percent level accuracy, with certain measurements considerably more accurate. A complete table listing these levels is presented for the first time in this thesis.

Additionally, a thorough discussion of the techniques used to record the positions of these levels is presented. Lineshapes associated with the common technique of photo*association* spectroscopy are discussed in great detail and compared to those associated with the less common photo*dissociation* spectroscopy. Careful evaluations of systematic effects are performed for a selection of levels to illustrate the capabilities and limitations of our experiment. Molecule-light coherence times approaching 10 ms are demonstrated, paving the way for future molecular clock studies.

Measurements of Zeeman Shifts (Chap. 4) When subjected to a magnetic field, the magnetic sublevels of a rovibrational level can split apart and shift via what's known as the "Zeeman effect." The magnitude of this shift can be related to the interaction of different types of angular momentum within the molecule, and can be a helpful tool for gaining more information about a molecule's structure.

1.3 This Thesis

We present detailed measurements of the linear Zeeman shifts for the majority of all observed levels in ^{88}Sr$_2$, most of which are made at the percent level or better. Fascinatingly, we observe certain rovibrational levels whose linear Zeeman shifts hew extremely closely to the values derived under the ideal Hund's case (c) approximation, and others which dramatically differ from this approximation. The fact that we can see both ideal and non-ideal behavior within the same molecule is explained as a consequence of whether or not Coriolis coupling with nearby levels is allowed or forbidden for different combinations of quantum numbers.

We also present tables of quadratic (and higher order) Zeeman shifts, and derive mathematical explanations for why the magnitude of the quadratic Zeeman shift increases approximately with the bond length to the power of $\frac{5}{2}$.

Finally, we describe the configuration of our magnetic Helmholtz coils, and show observable consequences of the $\sim$5 mV quantization of our DAQ-supplied control voltage.

Magnetic Control of Transition Strengths (Chap. 5) Electric dipole selection rules require that E1 transitions must connect states of opposite parity ($u \leftrightarrow g$ and $g \leftrightarrow u$) and $\Delta J = 0, \pm 1$. However, these rules can be broken in the presence of magnetic fields, which can cause mixing among nearby levels and cause previously "good" quantum numbers to become "bad."

We provide a simple framework for understanding this phenomenon based on perturbation theory, and discuss its experimental implications. This framework implies that previously forbidden transitions can become allowed in the presence of small magnetic fields. Specifically, the strengths of "singly-forbidden" transitions should increase quadratically with magnetic field, while the strengths of "doubly-forbidden" transitions should increase quartically.

We test these predictions by accurately measuring relative transition strengths for a series of "forbidden" transitions in ^{88}Sr$_2$. This study required developing state-of-the-art techniques for the quantitative measurement of transition strengths, which is surprisingly poorly described in the literature. We demonstrate a series of interesting effects, including observation of mixed quantization for transitions between states defined by orthogonal quantum axes and millionfold enhancement of the strengths of "forbidden" $\Delta J = 2, 3$ transitions with the application of magnetic fields of only a few tens of Gauss. We also discuss the relative strengths and weaknesses of three different techniques for quantitatively determining transition strengths.

Subradiant Spectroscopy (Chap. 6) We have observed several singly electronically excited "subradiant" states in ^{88}Sr$_2$, so-called because electric dipole radiative decay to the ground state is forbidden. These subradiant states are extremely long-lived, in some cases possessing lifetimes several hundreds of times longer than that of the $^3P_1$ state of ^{88}Sr. We precisely measure these lifetimes, and achieve record molecule-light coherence times. We show how the lifetimes of these states depend on bond length and magnetic field, and provide theoretical motivation for these behaviors. We also discuss in detail the experimental methods used to accurately measure these lifetimes, as well as for characterizing higher order M1 and E2 transition strengths from the ground state.

Carrier Thermometry in Optical Lattices (Chap. 7) Access to extremely narrow transitions to subradiant states has enabled our group to make detailed investigations of spectroscopic lineshapes not limited by the natural lifetimes of the final states. We derive accurate expressions for the lineshapes of transitions of particles confined to non-magic harmonic traps, and show why these expressions remain exact even in the presence of fourth-order corrections to the harmonic potential.

With knowledge of the lineshape's functional form, we invert the problem and describe a technique for *directly measuring molecular temperature* by fitting a spectrum with this lineshape. Previously, it has been impossible to directly measure the temperature of trapped, ultracold molecules lacking cycling transitions. We also demonstrate the first-ever observation of lattice sidebands in trapped ultracold molecules, and compare the temperatures derived from our lineshape-fitting technique with the more familiar process of comparing red and blue sideband areas. We use our new technique to investigate sources of heating in our molecular sample, and make the surprising discovery that in our experiment, molecules are hotter than the atoms from which they were photoassociated by more than a factor of two.

Finally, we discuss techniques for achieving high contrast, low-noise spectroscopic traces, including the removal of "cavity drift" in post-processing.

Photodissociation and Ultracold Chemistry (Chap. 8) When a molecule is subjected to sufficiently energetic laser light, it can break apart into fragments via a process called photodissociation. While this process has been known to and exploited by chemists for decades, it has received comparatively little attention from the precision measurements and ultracold molecules communities.

We demonstrate a series of experiments involving the photodissociation of ultracold molecules placed in well-defined quantum states with all quantum numbers controlled. We describe how information encoded in the angular distribution of the photofragments can reveal phenomena such as quantum interference and barrier tunneling. Finally, we list several unresolved mysteries which have the potential to better our understanding of how photochemistry behaves in the ultracold regime.

The work described in this chapter can also be found in our recently published article in Nature [4], as well as in the Master's Thesis of Florian Apfelbeck [1].

References

1. Apfelbeck, F.: Photodissociation dynamics of ultracold strontium dimers. Master's Thesis, Ludwig Maximilian University of Munich, Germany (2015)
2. Chin, C., Flambaum, V.: Enhanced sensitivity to fundamental constants in ultracold atomic and molecular systems near Feshbach resonances. Phys. Rev. Lett. **96**(23), 230801 (2006)
3. McDonald, M., McGuyer, B., Iwata, G., Zelevinsky, T. Thermometry via light shifts in optical lattices. Phys. Rev. Lett. **114**(2), 023001 (2015)
4. McDonald, M., McGuyer, B., Apfelbeck, F., Lee, C.-H., Majewska, I., Moszynski, R., Zelevinsky, T.: Photodissociation of ultracold diatomic strontium molecules with quantum state control. Nature **534**(7610), 122–126 (2016)

References

5. McGuyer, B., Osborn, C., McDonald, M., Reinaudi, G., Skomorowski, W., Moszynski, R., Zelevinsky, T.: Nonadiabatic effects in ultracold molecules via anomalous linear and quadratic Zeeman shifts. Phys. Rev. Lett. **111**(24), 243003 (2013)
6. McGuyer, B., McDonald, M., Iwata, G., Skomorowski, W., Moszynski, R., Zelevinsky, T.: Control of optical transitions with magnetic fields in weakly bound molecules. Phys. Rev. Lett. **115**(5), 053001 (2015)
7. McGuyer, B., McDonald, M., Iwata, G., Tarallo, M., Skomorowski, W., Moszynski, R., Zelevinsky, T.: Precise study of asymptotic physics with subradiant ultracold molecules. Nat. Phys. **11**(1), 32–36 (2015)
8. Reinaudi, G., Osborn, C., McDonald, M., Kotochigova, S., Zelevinsky, T.: Optical production of stable ultracold $^{88}Sr_2$ molecules. Phys. Rev. Lett. **109**, 115303 (2012)
9. Takano, T., Takamoto, M., Ushijima, I., Ohmae, N., Akatsuka, T., Yamaguchi, A., Kuroishi, Y., Munekane, H., Miyahara, B., Katori, H.: Real-time geopotentiometry with synchronously linked optical lattice clocks. arXiv preprint, arXiv:1608.07650v1 (2016)
10. Zelevinsky, T., Kotochigova, S., Ye, J.: Precision test of mass-ratio variations with lattice-confined ultracold molecules. Phys. Rev. Lett. **100**(4), 043201 (2008)

Chapter 2
Structure of ^{88}Sr and ^{88}Sr$_2$

Molecules are *complicated*, much more so than atoms. As Art Schawlow famously once said, "a diatomic molecule is a molecule with one atom too many." This complexity over atomic structure arises due to the new degrees of freedom available to molecules in the form of vibration and rotation, which causes the energy spectra for even relatively simple homonuclear diatomic molecules to become fabulously complex. In order to make sense of this chaos, physicists like to make simplifying approximations about molecular structure so that different species of molecules can be discussed using a common language.

Unfortunately, learning this common language is much like learning any language, in that you really need to absorb through immersion and osmosis: targeted questions can only take you so far. One difficulty is that some of the most important, seminal work laying the foundations for characterizing molecular structure was written in the 1920s and 1930s...and in German [4, 7]. The great, synthesizing textbooks [1–3] cite these important papers and work out special cases, and the best strategy for understanding molecular structure is to absorb these books and the relevant papers they cite. It might perhaps also be a good idea to learn German...

It would be foolish to attempt in one chapter of an experimental thesis to reproduce a body of knowledge which in reality takes a lifetime to master. Instead, this chapter will specifically focus on the structural features of ^{88}Sr$_2$. Specifically, a brief overview of the physics responsible for deciding which quantum numbers are "good" will be presented, as well as a discussion of the symmetries which restrict the set of allowed quantum numbers in both the electronic ground and excited states.

2.1 Rotational Structure

In theory, the full energy structure of a molecule can be ascertained by writing down its Hamiltonian and then solving the Schroedinger equation. We can start the process of simplification by first recognizing that the Hamiltonian $\hat{H}$ for a diatomic molecule can be divided into three different parts:

$$\hat{H} = \hat{H}_e + \hat{H}_v + \hat{H}_R, \tag{2.1}$$

where $\hat{H}_e$, $\hat{H}_v$, and $\hat{H}_R$ denote the electronic, vibrational, and rotational degrees of freedom of the molecule, respectively [6].

These three terms are fairly well decoupled from one another. The electronic energy of the molecule can be approximated as the total electronic energies of the atomic states forming the molecule (either $^1S_0$ or $^3P_1$ for the molecules considered in this thesis), and the vibrational energy can be characterized by a single number describing how quickly the nuclei vibrate with respect to one another (e.g., v can take any value between 1 and 62 for $^{88}Sr_2$ in the electronic ground state). The rotational part of the Hamiltonian is most complicated, however, because of the many different forms of angular momentum (spin and orbital, nuclear and electronic) which must be accounted for and properly added together to produce a total rotational energy. It is consideration of the rotational part of the Hamiltonian which will influence our choice for how to properly label the rovibrational levels of $^{88}Sr_2$.

The rotational Hamiltonian can be written in terms of the total rotational angular momentum $\hat{R}$ in the following way:

$$\hat{H}_R = B\hat{R}^2, \tag{2.2}$$

where the rotational constant $B = \frac{\hbar^2}{2\mu R^2}$ is related to the size (i.e., bond length) of the molecule R and its reduced mass μ, and will generally be a function of the vibrational state. However, the Hamiltonian above glosses over the fact that electronic momentum $\hat{J}_a = \hat{L} + \hat{S}$ can be carried by one or both of the Sr atoms, where $\hat{L}$ and $\hat{S}$ are the total orbital and spin electronic angular momenta, respectively. Since we are interested in how these various angular momenta interact, we should rewrite Eq. (2.2) as

$$\hat{H}_R = B\hat{R}^2 = B(\hat{J} - \hat{L} - \hat{S})^2, \tag{2.3}$$

where $\hat{J} = \hat{R} + \hat{J}_a$ is the total (rotational plus electronic) angular momentum of the molecule (excluding nuclear spin in this case because the ^{88}Sr nucleus is spinless).

Determining how to summarize the rotational energy of our molecule depends on which terms in the above Hamiltonian are most important, which is a complicated question to answer. Whether or not spin and orbital angular momentum can be

2.1 Rotational Structure

considered separately, and how strongly they couple to the total rotational angular momentum, will depend upon the size and structure of the atoms comprising a particular molecule.

2.1.1 Hund's Cases

The standard language used for characterizing different coupling types is to classify molecules according to different *Hund's cases*, which assign "good" quantum numbers based upon the relative strengths of coupling between different angular momenta. The $^{88}\text{Sr}_2$ molecule is best described by either Hund's case (a) or (c), depending on whether we are discussing the electronic ground state or excited state. An excellent description of when and why various Hund's cases apply in different situations is given by Stepanov and Zhilinskii [5]. Here we'll just make a few brief remarks.

2.1.1.1 Hund's Case (a)

In Hund's case (a), it is assumed that the orbital angular momentum $\hat{L}$ is strongly coupled to the internuclear axis, while the electronic spin $\hat{S}$ is strongly coupled to $\hat{L}$ [2]. The result is a situation in which we have a maximal number of "good" quantum numbers:

- Λ, the projection of the electronic orbital angular momentum $\hat{L}$ onto the internuclear axis.
- Σ, the projection of the electronic spin angular momentum $\hat{S}$ onto the internuclear axis.
- S, the total electronic spin angular momentum of the system.
- J, the total angular momentum (rotational plus electronic).
- Ω, the projection of the electronic angular momentum $\hat{L}+\hat{S}$ onto the internuclear axis.

Hund's case (a) is a good description of the electronic ground state of $^{88}\text{Sr}_2$, which possesses no electronic angular momentum at all. However, it turns out that strontium molecules comprised of a ground state atom plus an excited atom will be better described by a different approximation.

2.1.1.2 Hund's Case (c)

In Hund's case (c), it is assumed that the spin-orbit coupling between $\hat{L}$ and $\hat{S}$ is stronger than that of either to the internuclear axis. In this case, the $\hat{L}$ and $\hat{S}$ operators combine to form a total electronic angular momentum operator $\hat{J}_a$ which is only weakly coupled to the rotational motion of the nuclei. Rewriting the rotational Hamiltonian as

$$\hat{H}_R = B(\hat{J} - \hat{J}_a)^2 = B(\hat{J}^2 + \hat{J}_a^2 - 2\hat{J} \cdot \hat{J}_a), \tag{2.4}$$

assuming Hund's case (c) is a valid description then amounts to assuming that the term $-2B\hat{J} \cdot \hat{J}_a$ is small [6]. The Hamiltonian then consists of two parts: a "zeroth-order" contribution $\hat{H}_R^0 = B(\hat{J}^2 + \hat{J}_a^2)$, which informs which labels we choose to describe our molecule's rotational levels; and a "perturbation" contribution $\hat{H}_R^1 = -2B\hat{J} \cdot \hat{J}_a$, which will cause mixing between the zeroth-order basis states.

The operators which commute with $\hat{H}_R^0$ should serve as the labels for our Hund's case (c) basis states, since their eigenfunctions will diagonalize the Hamiltonian. Upon inspection, it's clear that $\hat{J}^2$, $\hat{J}_a^2$, and their projections will commute with $\hat{H}_R^0$, and so we will label our Hund's case (c) basis states with four quantum numbers J, J_a, M_J, and Ω, as well as a catch-all label $\eta(\Omega)$ describing electronic and vibrational degrees of freedom (itself labeled by Ω for purposes of bookkeeping):

$$|\Psi\rangle_{\text{Hund's case (c)}} = |\eta(\Omega), J_a; J, M_J, \Omega\rangle \tag{2.5}$$

A semicolon separates the labels J_a and J to indicate that these two vector spaces are decoupled from one another. This assumption amounts to requiring the perturbation $\hat{H}_R^1 = -2B\hat{J} \cdot \hat{J}_a$ to be small. Cases for which $\hat{H}_R^1$ is not small result in *Coriolis coupling*, whereby states with different Ω can be mixed together. See Sect. 4.2.2.2 for details.

Given this set of possible quantum numbers to work with, let's now consider what possibilities are allowed for ^{88}Sr$_2$ molecules in both the electronic ground and singly-excited ^{1}S$_0$+^{3}P$_1$ states.

2.1.2 Electronic Ground State

In the ^{1}S$_0$+^{1}S$_0$ electronic ground state, the situation is relatively simple. Neither component atom of the strontium molecule carries either spin or orbital angular momentum (i.e., $L = S = 0$), implying that both the total electronic angular momentum J_a and its projection along the internuclear axis Ω must equal 0.

We can also make some general statements about the symmetry required of such a molecule. Because the ^{88}Sr nucleus is bosonic, it *must* be true that upon exchange of the two nuclei, the total molecular wavefunction should retain the same sign.

However, whether or not the molecular wavefunction acquires a minus sign can also be determined from the symmetries of the electronic wavefunctions and the molecular wavefunction's quantum numbers. According to Herzberg (Section V,2c, p. 238) [3], nuclear exchange for a molecular wavefunction comprised of two even, electronic ground state atomic wavefunctions will remain unchanged for *even* values of J, and acquire a minus sign for *odd* values of J. (Why exactly this is so is *complicated*. A hint is that the nuclear exchange operator is equivalent to a reflection of all particles [electrons plus nuclei] at the origin, followed by a reflection of only the electrons at the origin. Each of these operations has a well-defined effect on the rotational, vibrational, and electronic wavefunctions whose product comprises the total molecular wavefunction. Determining the sign behavior of nuclear exchange is then a matter of evaluating the effect of each inversion on each component of the wavefunction.)

This result leads to dramatic consequences. If odd-J states acquire a minus sign upon nuclear exchange, but bosonic nuclei require that the wavefunction remain unchanged, then it must be true that *only even J levels are allowed in the electronic ground state*. Sure enough, we have so far only observed even-J in the ground state. See Table 3.2 for details.

There is one additional symmetry which must be true of ground state molecules. The inversion symmetry of the total ground state molecular wavefunction must be *even* since it is composed of two atoms in identical states. Odd symmetry is impossible because symmetrization would force the wavefunction to equal zero. This can also be thought of as a result of the Wigner-Witmer rules [3, 7]. The name we give this inversion symmetry is *gerade* for even and *ungerade* for odd.

2.1.3 Electronic Excited State

The electronic state is naturally more complicated. First, because the atomic wavefunctions of each of the component atoms are different, the molecular wavefunction can have either *ungerade* or *gerade* symmetry. However, bosonic symmetry upon exchange of the nuclei must still be respected.

Because the $^3P_1$ atom carries 1 unit of electronic angular momentum J_a, the total projection Ω of $\hat{J}_a$ onto the internuclear axis can take on the value of either 0 or 1. This leads to four distinct possible combinations of inversion symmetry ($= u, g$) and Ω ($= 0, 1$) which serve as labels for our singly excited rovibrational levels: 1_u, 0_u, 1_g, and 0_g. And by considering once again how each component of the wavefunction transforms under different symmetry rules, we arrive at the following restrictions for the quantum numbers of singly-excited rovibrational levels:

- 1_u supports $J \geq 1$
- 0_u supports $J \geq 1$ and *odd*
- 1_g supports $J \geq 1$
- 0_g supports $J \geq 0$ and *even*

These restrictions on the possible combinations of J, g/u, and Ω will have dramatic implications for both the interpretation of linear Zeeman shifts and our ability to control all quantum numbers in both the initial and final state of photodissociation experiments. See Fig. 2.1 for an illustration of the various potentials describing $^{88}Sr_2$ molecules in the electronic ground and singly excited states.

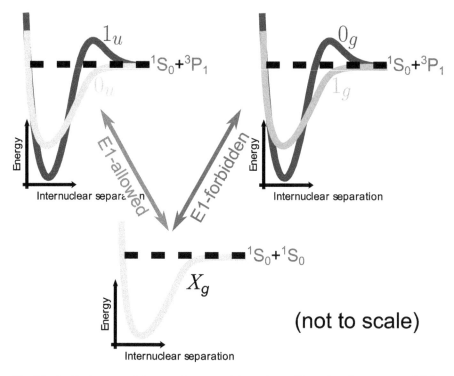

Fig. 2.1 Qualitative potential curves (not to scale) are drawn illustrating the structure of $^{88}\text{Sr}_2$ molecules in both the electronic ground and singly excited states for different values of inversion symmetry and Ω. E1 transitions connect states of opposite inversion symmetry and $\Delta J = 0, \pm 1$. E1, M1, and E2 selection rules combined with the restriction of J to either even or odd values for different potentials enable observations of pure Hund's case (c) linear Zeeman shifts (Chap. 4), precise measurements of strongly forbidden transition strengths (Chap. 5), precise determinations of transition strengths to and lifetimes of highly subradiant states (Chap. 6), and explorations of quantum state resolved photochemistry (Chap. 8 and attached publication)

2.1.3.1 Aside: On Parity-Adapted Wavefunctions

In Hund's case (a) and (c), the rotational energy for a molecule with total rotational angular momentum J and projection along the internuclear axis Ω is given by the following expression [3]:

$$E_{\text{rot}} = B_v [J(J+1) - \Omega^2], \tag{2.6}$$

where B_v is a vibrational level-dependent rotational constant. What's interesting about this equation is that *rotational energy doesn't depend on the sign of Ω*. This means that $\Omega = \pm 1$ are energy-degenerate. Therefore to truly diagonalize our Hamiltonian, we should use parity-adapted superpositions of positive and negative Ω [8]:

$$|\Psi\rangle_{\text{Hund's case (c)}} = \frac{1}{\sqrt{2}}\big(|\eta(\Omega),J_a;J,M_J,\Omega\rangle + (-1)^p|\eta(-\Omega),J_a;J,M_J,-\Omega\rangle\big), \tag{2.7}$$

where p is the parity of the state being considered. This fact will have implications for how we interpret the state mixing responsible for linear Zeeman shifts of 0_u levels, and may additionally have something to do with the mystery of magnetic field dependence of the $1_u/0_u$ potential alluded to in Chap. 8.

References

1. Atkins, P., Friedman, R.: Molecular Quantum Mechanics. Oxford University Press, Oxford (2011)
2. Brown, J., Carrington, A.: Rotational Spectroscopy of Diatomic Molecules. Cambridge University Press, Cambridge (2003)
3. Herzberg, G. *Molecular Spectra and Molecular Structure I. Spectra of Diatomic Molecules*. Princeton: D Van Nostrand Company, Inc., 1950
4. Hund, F.: Allgemeine Quantenmechanik des Atom- und Molekelbaues. Handbuch der Physik, vol. 24, p. 561. Springer, Berlin (1933)
5. Stepanov, N., Zhilinskii, B.: When and why Hund's cases arise. J. Mol. Spectrosc. **52**(2), 277–286 (1974)
6. Veseth, L.: Hund's coupling case (c) in diatomic molecules. I. Theory. J. Phys. B At. Mol. Phys. **6**(8), 1473 (1973)
7. Wigner, E., Witmer, E.: Über die Struktur der zweiatomigen Molekelspektren nach der Quantenmechanik. Z. Phys. **51**, 859 (1928)
8. Xie, J., Zare, R.: Selection rules for the photoionization of diatomic molecules. J. Chem. Phys. **93**(5), 3033–3038 (1990)

Chapter 3
Measurements of Binding Energies

The first step toward understanding a molecule's structure is to measure its spectrum, i.e. the locations of its energy levels. A careful analysis of this spectrum can provide a wealth of knowledge which can then be applied toward numerous applications. One current application lies in the design of direct laser-cooling schemes for molecules, so as to "engineer" cycling transitions which can accommodate a photon scattering rate sufficient for producing a large cooling force [27]. Another, more germane to the work described in this thesis, is related to searches for new physics.

For a simple molecule such as ^{88}Sr$_2$, amenable both to ab initio theoretical modeling and ultra-high precision spectroscopy, the energy spectrum provides a benchmark which can be compared with a high level of accuracy to the predictions of state-of-the-art quantum chemistry models. In this scenario, measurements of the spectrum can serve as valuable feedback for the improvement of these models, which as a result have improved dramatically in recent years [3, 26].

With a better understanding of quantum chemistry in hand, we're particularly interested in combining those theoretical tools with the measurement precision achievable with spectroscopy in an optical lattice to probe subtle effects such as quantum electrodynamics (QED), as well as physics beyond the current Standard Model. We have in mind several future experiments which might accomplish this goal.

One possibility is a careful measurement of the differences in binding energies between weakly, intermediately, and deeply bound molecules in the electronic ground state over a long ($\sim$1 year) period of time. Because the *vibrational* level structure depends strongly on how heavy the nuclei are compared to their electron clouds, and because variations in that mass ratio would affect intermediately bound levels more strongly than deeply or weakly bound levels, such a self-referencing "molecular clock" measurement would allow for constraints to be placed on the possible time variation of the electron to proton mass ratio in a way which would be only weakly model-dependent [4, 31].

Another exciting experiment would involve making precise measurements of binding energies for very deeply bound molecules in the same rovibrational state, but comparing among the six stable even isotopologues (^{88}Sr$_2$, ^{86}Sr$_2$, ^{84}Sr$_2$, ^{88}Sr^{86}Sr, etc.). With an accurate enough understanding of the isotope shifts in Sr$_2$, one could begin to look for deviations from theory, and attempt to assign these deviations to "new physics." One source of new physics might be an anomalous gravitational interaction between the nuclei. If the gravitational attraction between two masses were to deviate strongly from Newtonian physics at small length scales, that deviation could manifest itself as a difference in the vibrational level spacing among molecules with different nuclear masses. Though such an effect would presumably be extremely small, the fact is that Newtonian gravity is quite poorly constrained at the few-nanometer scale characterizing the bond lengths of strontium dimers [10]. Such an experiment would rely heavily on precise modeling of the internuclear potential for strontium and a thorough understanding of isotope shifts [12], which is excellent motivation for working to perfect our understanding of quantum chemistry [25].

In this chapter I will describe several techniques we have adapted for precise measurements of binding energies of molecules in both ground and excited electronic states. In approximate order from low spectroscopic resolution to high, they are: *photoassociation spectroscopy*, involving the conversion of free atoms into molecules; *photodissociation spectroscopy*, involving the fragmentation of a molecule into energetic atoms; and *bound-bound spectroscopy*, involving transitions between bound states. Perfecting techniques such as these will be essential in designing new molecular spectroscopy experiments in the future.

3.1 Catalog of All Known Rovibrational Levels

Besides possibilities of making precision tests of fundamental physics with targeted measurements of certain levels, it is also true that with access to large data sets comes the possibility of discovering deeper patterns and broader generalities about the physics of the system in question. For that reason, we present here a summary of our lab's 6-year effort to characterize the energies of nearly all weakly bound levels near the singly excited $^1S_0+^3P_1$ dissociation threshold of ^{88}Sr$_2$, as well as improved values for several of the levels near the ground state $^1S_0+^1S_0$ threshold. The availability of accurate binding energies for the $J = 1$ levels has already led to a better understanding of Coriolis coupling and of the quantum chemistry required to predict molecular structure [3], and it's exciting to imagine what new progress, e.g. in molecular QED, could be made with the complete list of the higher J states presented here. And even without having to appeal to the precise comparison of esoteric computations to accurate data, when looking at these tables one can discern certain patterns that hint toward a deeper level of understanding. For example, comparing the level spacings among the 1_u, 1_g, and 0_u potentials yields a hint about their relative shapes which can be grasped immediately without having to rely on

3.1 Catalog of All Known Rovibrational Levels

complex calculations or exact measurements. The fact that only odd J are possible for the 0_u states hints at the importance of quantum statistics in determining the structure of molecules (see Chap. 2). And a careful look at the rotational spacing of different states reveals that it gets *larger* for molecules which are more *deeply bound*—exactly the result we'd expect from a semiclassical picture of a molecule which gets *smaller* when it is trapped more deeply within the well.

3.1.1 $^1S_0 + {}^3P_1$

Table 3.1 lists the measured binding energies of all currently known rovibrational levels of singly excited ($^1S_0 + {}^3P_1$) $^{88}Sr_2$. This represents a complete accounting of all rovibrational levels with $J \leq 4$ and $E_b < 8.5$ GHz, with the exception of levels within the 0_g manifold, which so far remain unobserved. The most accurate binding energy measurements are of the $1_g(-1, 1)$, $0_u(-1, 1)$, and $0_u(-2, 1)$ states, which are known to few-kHz accuracy. For these states, careful evaluations of systematic shifts have been performed. Determination of the absolute binding energies of these levels was limited to a few kHz due to the difficulty in determining the exact location of the broad dissociation threshold (∼15 kHz width). For states with linewidths much narrower than 15 kHz, such as those within the 1_g manifold, relative binding energy differences between very narrow states could in principle be determined much more accurately, since relative binding energy differences would not be tied to the broad "shelf" lineshape of the dissociation threshold.

Many of the listed states cite uncertainties of 0.5 or 1 MHz. These are conservative estimates, acknowledging that no systematic evaluations of the effects of light shifts, magnetic fields, etc. have been performed. But there is no inherent roadblock (besides an investment of experiment time) to determining the binding energies of these states to a much greater precision in the future.

3.1.2 $^1S_0 + {}^1S_0$

Table 3.2 lists the binding energies of all currently known rovibrational levels in the electronic ground state of $^{88}Sr_2$. Upon inspection, two things are immediately clear. One is that the electronic ground state is far simpler than the electronic excited state. The lack of electronic angular momentum in the constituent atoms means that there's only one possibility for its projection along the internuclear axis (namely 0), which further restricts the possible symmetry of the wavefunction to *gerade* and the values of the rotational angular momentum J to be *even* (see Chap. 2). Another is that our knowledge of the ground state is less complete than that of the excited state, with only three vibrational manifolds found and J only as large as 2. Precise measurements of more deeply bound vibrational levels will require phase-locking lasers with a frequency comb, since phase offset-locking is restricted to $\lesssim 9$ GHz

Table 3.1 A complete list of the most accurate binging energy measurements for all singly excited ($^1S_0+{}^3P_1$) rovibrational states of $^{88}Sr_2$ with $E_b <$ 8.5 GHz and $J \leq 4$, neglecting only states within the 0_g manifold (which so far remain unobserved)

	$J=1$		$J=2$		$J=3$		$J=4$	
	Energy	Source	Energy	Source	Energy	Source	Energy	Source
$0_u(-1,J)$	0.4653(45)	McDonald [14]	X		X		X	
$0_u(-2,J)$	23.9684(50)	McGuyer et al. [19]	X		0.626(12)	McGuyer et al. [19]	X	
$0_u(-3,J)$	222.161(35)	Zelevinsky et al. [30]	X		131.9(5)*	McDonald [14]	X	
$0_u(-4,J)$	1084.093(33)	Zelevinsky et al. [30]	X		901.0(5)*	McDonald [14]	X	
$0_u(-5,J)$	3463.280(33)	Zelevinsky et al. [30]	X		3179.1(5)*	McDonald [14]	X	
$0_u(-6,J)$	8429.650(42)	Zelevinsky et al. [30]	X		8076.1(5)*	McDonald [14]	X	
$1_u(-1,J)$	353.236(35)	Zelevinsky et al. [30]	287(1)	McGuyer et al. [18]	171(1)*	McDonald [14]	56.2(1.0)	McGuyer et al. [18]
$1_u(-2,J)$	2683.722(32)	Zelevinsky et al. [30]	2569(1)*	McDonald [14]	2355.2(5)*	McDonald [14]	2154(1)*	McDonald [14]
$1_u(-3,J)$	8200.163(39)	Zelevinsky et al. [30]	8113(1)*	McDonald [14]	7660.6(5)*	McDonald [14]	7467.5(1.0)*	McDonald [14]
$1_g(-1,J)$	19.0420(38)	McGuyer et al. [20]	7(1)	McGuyer et al. [20]	X		X	
$1_g(-1,J)$	316(1)	McGuyer et al. [20]	270(1)	McGuyer et al. [20]	193(1)	McGuyer et al. [20]	114(1)*	McDonald [14]
$1_g(-1,J)$	1669(1)	McGuyer et al. [20]	1581(1)	McGuyer et al. [20]	1438(1)	McGuyer et al. [20]	1271(1)*	McDonald [14]
$1_g(-1,J)$	5168(1)	McGuyer et al. [20]	5035(1)	McGuyer et al. [20]	4826(1)	McGuyer et al. [20]	4570(1)*	McDonald [14]

"Energy" describes binding energy defined with respect to the dissociation threshold (in MHz), and "Source" describes the reference from which the cited binding energy is taken. Note that entries marked with "*" represent rovibrational levels which were previously unreported, and therefore represent discoveries unique to this thesis. Entries marked with "X" refer to rovibrational levels which are forbidden by either selection rules or energy considerations

3.2 Photo*association* (Free-Bound) Spectroscopy

Table 3.2 A list of all currently known electronic ground state levels of ^{88}Sr$_2$

	$J=0$		$J=2$	
	Energy	Source	Energy	Source
$X(-1,J)$	136.6447(50)	McDonald [14]	66.6(2)	Martinez de Escobar et al. [13]
$X(-2,J)$	1400.1(2)	Reinaudi et al. [24]	1245.6(2)	McGuyer et al. [19]
$X(-3,J)$	5110.6(1)	Reinaudi et al. [24]	?	

Note that the uncertainty in the binding energy of the $X(-1,0)$ state reported here is 5 kHz, representing an improvement by a factor of 40 over previously reported measurements [13]

due to RF electronics limitations in our experiment. And the reason we have only observed ground state levels with $J = 0, 2$ is because only those states are produced naturally via spontaneous decay from the $J = 1$ electronically excited levels to which we can photoassociate.

3.2 Photo*association* (Free-Bound) Spectroscopy

When we say that a pair of atoms has been *photoassociated*, we mean that they have absorbed a photon and subsequently bound together into a molecule. Photoassociation spectroscopy [9] is a common tool for studying molecular spectra, and has been used previously to characterize the binding energies of electronic excited states and electronic ground states in ^{88}Sr$_2$ to few-tens and few-hundreds of kHz precision, respectively [13, 30].

In our experiment we are particularly interested in using photoassociation (PA) as a tool for the creation of ground state molecules. To do so efficiently requires being able to characterize how strongly a particular molecular electronic excited state will decay into a particular electronic ground state. This can be accomplished by observing the splitting of a PA peak into an *Autler-Townes doublet*, but to do so accurately requires understanding precisely the lineshape characterizing this process.

In this section I will describe our work to better understand the photoassociation lineshape under various experimental conditions, and discuss the physics we've extracted through applying this understanding to measurements.

3.2.1 Observing Photoassociation Losses

When a single laser is used to coerce pairs of atoms in the electronic ground state ($^1S_0+^1S_0$) into a molecule in the electronic excited state ($^1S_0+^3P_1$), the resulting molecule will be unstable, and will therefore rapidly decay. For all except those most weakly bound (which can decay directly into free atoms), these molecules will primarily decay into bound molecules in the electronic ground state. Similarly, two-photon transitions will either directly produce molecules in the electronic ground state, or molecules which decay quickly from unstable excited states. In either case, for an experiment (such as ours) which counts the number of *atoms* remaining in a cloud via absorption imaging on a strong atomic cycling transition, these decay pathways will result in *dark states*, invisible to our imaging scheme. By sweeping a laser across a photoassociation transition and counting atom losses, we can therefore obtain information about the molecular excited state in question.

3.2.1.1 Considerations for Determining the Lineshape

The rate at which atoms are photoassociated will depend on their density *squared*, since one atom must "find" another in order for photoassociation to occur. The differential equation describing the density of an atom cloud whose losses are dominated by these two-body, density-dependent losses ("PA losses"), and also incorporating one-body losses due to, e.g., heating by the trap laser ("vacuum losses"), can be written as:

$$\frac{dn(\delta)}{dt} = -2K_{\text{eff}}(\delta) \cdot n(\delta)^2 - \Gamma \cdot n(\delta), \tag{3.1}$$

where δ is the detuning(s) of the photoassociation laser(s) from resonance, $n(\delta)$ is the density of atoms in the trap, Γ is the one-body "vacuum" loss rate, and $K_{\text{eff}}(\delta)$ is an "effective" photoassociation rate governed by the rate at which collisions between pairs of atoms occur. This effective rate takes into account an integration over all possible relative collision energies within the gas, as will be discussed later.

There is a technical detail in an experiment like ours in that we don't have experimental access to the atom density n, but rather only the total atom number N. We can make an approximation, however, by substituting into Eq. (3.1) $n = A \cdot N$, i.e. by assuming that density is proportional to the atom number [23]. This approximation is exact in the limit of a uniform density across the cloud, which obviously isn't exactly true in an optical lattice. For the purposes of this analysis, however, we'll ignore this detail. (Note however that this assumption represents a critical difference between photo*association* and photo*dissociation* spectrosocopy which will be elaborated upon later.) Making this substitution yields the following equation:

$$\frac{dN(\delta)}{dt} = -2A \cdot K_{\text{eff}}(\delta) \cdot N(\delta)^2 - \Gamma \cdot N(\delta), \tag{3.2}$$

This equation has the following exact solution, describing atom losses after a photoassociation pulse of length τ:

$$N_\tau(\delta) = \frac{N_0 e^{-\Gamma \tau}}{1 + \frac{2A \cdot K_{\text{eff}}(\delta) \cdot N_0}{\Gamma}(1 - e^{-\Gamma \tau})}. \tag{3.3}$$

3.2 Photo*association* (Free-Bound) Spectroscopy

Vacuum losses in our experiment are small on the timescale of the $\lesssim 100$ ms PA pulse [23, 24]. Taking the limit of Eq. (3.3) as $(\Gamma \tau) \to 0$ yields:

$$N_\tau(\delta) = \frac{N_0}{1 + 2A \cdot K_{\text{eff}}(\delta) \cdot N_0 \cdot \tau}, \tag{3.4}$$

where A is in our experiment left as an empirically determined fit parameter.

The "Effective" PA Rate $K_{\text{eff}}(\delta)$ The photoassociation rate describes the likelihood that two atoms will collide to form a molecule when brought close to one another in the presence of laser light. In fact, this probability should also depend upon the relative collision energy of the atoms, since conservation of energy must be satisfied such that the sum of the energy of the photoassociation laser(s) and the kinetic energy of the colliding atom pair combines to yield the total energy of final molecule, which will equal the electronic energy of the excited $^3P_1$ atom minus the molecular binding energy.

We incorporate this energy-dependence by integrating an energy-dependent PA rate $K_\epsilon(\delta)$ over all possible collision energies. (Quantum mechanically, we are actually calculating a total transition rate by summing over all possible output channels.) However, because the atoms in the cloud being probed are thermally distributed among many energies, we must weight each $K_\epsilon(\delta)$ by the likelihood that a particular energy ϵ will be represented. The end result is the following equation:

$$K_{\text{eff}}(\delta) = \frac{1}{Z} \int_V \int_0^\infty K_\epsilon(\delta) e^{-\frac{\epsilon}{kT}} g(\epsilon) dV d\epsilon, \tag{3.5}$$

where Z is the partition function, $g(\epsilon)$ is the density of states, and the integral over volume V is for bookkeeping purposes and will disappear in this approximation (since we are assuming uniform trap density) [30]. The exact form of $K_\epsilon(\delta)$ can be calculated quantum mechanically, and will depend upon whether one or two lasers participate in the photoassociation process [2, 22].

Dimensionality Considerations In our experiment, the atoms are tightly confined to a 1D optical lattice with a small (spectroscopically unresolved) radial trap frequency, ensuring that collisions primarily occur in directions transverse to the trapping axis with an approximately continuous distribution of collision energies. Building this physics into Eq. (3.5) means summing over a continuous energy distribution in two dimensions, i.e. using $g_{2D}(\epsilon)d\epsilon = \frac{m}{\pi\hbar^2}d\epsilon$ and $Z_{2D} \equiv \int_V \int_0^\infty g_{2D}(\epsilon)e^{-\frac{\epsilon}{kT}}dVd\epsilon = \frac{m}{\pi\hbar^2}(kT)V_{2D}$, which gives the following:

$$K_{\text{eff}}^{2D}(\delta) = \int_0^\infty K_\epsilon(\delta) e^{-\frac{\epsilon}{kT}} \frac{d\epsilon}{(kT)}. \tag{3.6}$$

However, though collisions in the axial direction should be suppressed due to the large axial trap spacing, they will not be completely negligible. Parity considerations will restrict interactions to only those between atoms separated by even multiples of the trap quantum $\hbar\omega_x$, where ω_x is the axial trap angular frequency of the lattice [19]. (Incidentally, this statement is also true for collisions in the transverse direction. However, because the radial trap frequency is so small, assuming a continuous energy distribution and ignoring the restriction to even multiples of the trap frequency leads to no experimentally observable differences.)

To account for the contributions of collisions in the axial direction, we can include in the above integral a weighted sum over axial trap states:

$$K_{\text{eff}}^{(2D+\text{axial})}(\delta) \propto \sum_{n=0}^{\infty} e^{-\frac{2n\hbar\omega_x}{kT}} \int_0^{\infty} K_{(\epsilon+2n\hbar\omega_x)}(\delta) e^{-\frac{\epsilon}{kT}} \frac{d\epsilon}{(kT)}. \tag{3.7}$$

By substituting Eq. (3.7) into Eq. (3.3), and using the correct expression for $K_{\text{eff}}^{(2D+\text{axial})}$ as a function of photoassociation laser frequencies (discussed in the following sections), we obtain a spectroscopic lineshape describing N_τ as a function of laser detuning.

3.2.2 One-Photon (Electronic Excited State)

For one-photon PA, $K_\epsilon(\delta_1)$ will have the form of a modified Lorentzian, where δ_1 is the detuning from resonance as defined in Fig. 3.1a [22]:

$$K_\epsilon(\delta_1) = C \cdot \frac{\gamma_s(\epsilon)}{(\epsilon/h + \delta_1 - \delta_{1c})^2 + \left(\frac{\gamma_b + \gamma_s(\epsilon)}{2}\right)^2}. \tag{3.8}$$

In the above equation, δ_{1c} is the location of the PA resonance for $\epsilon = 0$, δ_1 is the detuning from resonance of the PA laser being swept across resonance, γ_b is an empirically determined broadening parameter accounting for our observed linewidths, $\gamma_s(\epsilon)$ is an energy-dependent linewidth representing decay probability to a continuum state with energy ϵ, and C is a scaling pre-factor accounting for Franck-Condon overlap between continuum and bound state [2]. In practice, this equation is more complicated than necessary. For the states for which photoassociation measurements will be covered in detail in this thesis, the decay probability to continuum is both very small and approximately constant across the few-μK energies characterizing our atom cloud. In this case we can make a good approximation by assuming γ_s to be constant (which additionally simplifies the math considerably), yielding our operational fitting function:

$$K_\epsilon(\delta_1) = C \cdot \frac{\gamma_s}{(\epsilon/h + \delta_1 - \delta_{1c})^2 + \left(\frac{\gamma_b + \gamma_s}{2}\right)^2}. \tag{3.9}$$

Combining Eq. (3.9) with Eqs. (3.7) and (3.4) yields a lineshape which is a function of six free parameters: T, the temperature of atomic gas; N_0, the initial

3.2 Photo*association* (Free-Bound) Spectroscopy

$$\Delta_1 \equiv \delta_1 - \delta_{1c} = E_{b1} - \hbar\omega_1$$
$$\Delta_2 \equiv \delta_{2c} - (\delta_1 - \delta_{1c}) = E_{b2} - \hbar(\omega_1 - \omega_2)$$

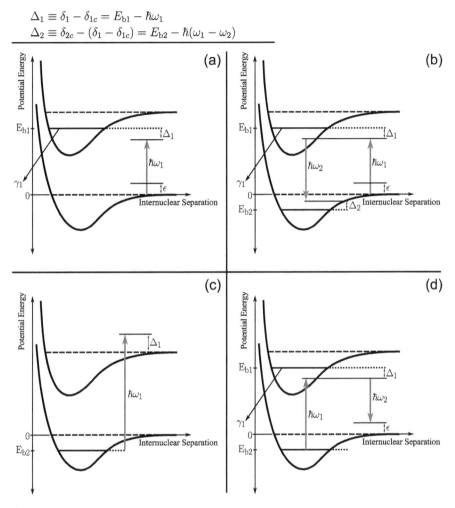

Fig. 3.1 Illustrations of laser detunings used for (**a**) one-photon photoassociation spectroscopy, (**b**) two-photon photoassociation spectroscopy, (**c**) one-photon photodissociation spectroscopy, and (**d**) two-photon photodissociation spectroscopy. Definitions of detunings in panels (**a**) and (**b**) are based on [22] and [2], respectively

(off-resonant) atom number; δ_{1c}, the location of the 1-photon PA resonance (which is determined by the spectroscopy laser's absolute calibration); $(\gamma_b + \gamma_s)$, a linewidth-broadening coefficient unrelated to thermal broadening; $(C \cdot \gamma_s \cdot A)$, a scaling pre-factor accounting for both the Franck-Condon overlap between ground and excited state as well as the proportionality constant between atom number and density in an optical lattice; and ω_x, the axial trap frequency of the 1D optical lattice. In principle, the axial trap frequency can be determined separately either

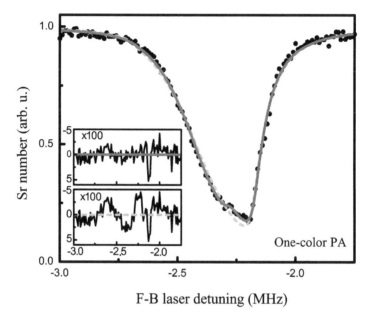

Fig. 3.2 This figure shows experimental data resulting from one-photon photoassociation into the $0_u(-4, 1)$ state. The data is fitted by combining Eq. (3.9) with either Eq. (3.6) (blue) or (3.7) (red). In the figure above, residuals for each fit are plotted in the inset boxes at the lower left corner. Note that the small "bump" on the left side of the trace is fitted well by Eq. (3.7), i.e. by accounting for discrete axial trap motion. This figure has been adapted from [19]

by monitoring atom losses induced by "lattice shaking" [7] or spectroscopically resolving sideband transitions [15]. However, in our experiment we generally left this as a free parameter as well.

Figure 3.2 shows a representative trace of a one-photon PA spectrum. Note that the bump on the left side of the PA lineshape is well represented by proper accounting for the effect of a quantized axial trap frequency [19].

3.2.3 Two-Photon (Electronic Ground State)

If a second laser, tuned close to resonance with a transition from a molecular level in the electronic ground state to one in the electronic excited state, is applied during the photoassociation pulse, the single photoassociation peak will split into two (see Fig. 3.1). When the second laser (L_{BB}) is far from resonance, we can think of the condition for losses in the PA spectrum as adhering to one of two scenarios: either (1) the first laser (L_{FB}) is tuned to resonance with the transition from free atoms to an electronic excited state, or (2) the difference in frequencies between L_{BB} and L_{FB} is equal to the binding energy of a level in the electronic ground state. As L_{BB} gets

3.2 Photoassociation (Free-Bound) Spectroscopy

closer and closer to resonance, this picture begins to break down: we can no longer assign each dip in the spectrum to the addressing of a particular rovibrational level, but rather must invoke a superposition of the two.

The splitting of a single PA line into a doublet is called the *Autler-Townes effect* [5]. Following Bohn and Julienne [2], the PA rate K in this case has the form:

$$K_\epsilon(\delta_1) = C \frac{\gamma_b \gamma_s (\epsilon/h - \Delta_2)^2}{[(\epsilon/h - \Delta_+)(\epsilon/h - \Delta_-)]^2 + \frac{1}{4}(\gamma_b + \gamma_s)^2 (\epsilon/h - \Delta_2)^2}, \quad (3.10)$$

where I have again assumed the decay rate γ_s to the atomic continuum to be small and constant across all relevant collision energies. In the equation above, several new variables have been introduced in addition to those present in Eq. (3.9):

- $\Delta_\pm = \frac{1}{2}(\Delta_1 + \Delta_2) \pm \frac{1}{2}\sqrt{(\Delta_1 - \Delta_2)^2 + 4\Omega_{12}^2}$
- $\Delta_1 = -(\delta_1 - \delta_{1c})$
- $\Delta_2 = \delta_{2c} - (\delta_1 - \delta_{1c})$

where δ_{2c} is detuning of L_{BB} from the bound-bound resonance and Ω_{12} is the "molecular Rabi coupling," or rather (operationally) the minimum frequency separation between the two peaks comprising the PA spectrum, occurring when L_{BB} is exactly on resonance.

Whereas Eq. (3.9) can be used to study rovibrational levels in the electronic excited state, Eq. (3.10) can be used to study rovibrational levels in the electronic ground state. In particular, binding energies can be determined by the frequency difference of lasers L_{BB} and L_{FB} when the splitting between two peaks in the Autler-Townes doublet is minimized.

3.2.3.1 Autler-Townes Spectroscopy

Figure 3.3 shows a representative set of two-photon PA traces interrogating the $0_u(-6, 1) \leftrightarrow X(-3, 0)$ transition. The binding energy of the ground state is determined by first taking several spectra. For each spectrum, the detuning of L_{BB} is changed by a discrete amount, and L_{FB} is then swept across resonance to reveal the locations of the two PA peaks. Figure 3.3a shows a plot of δ_{2c} vs the detuning of L_{BB}, which reveals a linear dependence with a slope of $+1$. (Any deviation would imply a disagreement with the model used for fitting the two-photon spectrum.) When L_{BB} is on resonance, the PA peaks will be symmetrically split into a doublet, and the fitted value of δ_{2c} will be zero. We can determine the value of L_{BB}'s detuning necessary to achieve this condition by fitting a line to Fig. 3.3a and calculating the x-intercept. Figure 3.3b shows δ_{1c} at several detunings of L_{BB}. Since δ_{1c} can be thought of as the "true" location of the 1-photon resonance, this value should be the same for all fits. The binding energy can then be determined by evaluating the difference in the frequencies of L_{BB} and L_{FB} when $\delta_{2c} = 0$.

If we plot the locations of the two photoassociation peaks Δ_+ and Δ_- vs the detuning of L_{BB} (Fig. 3.3c), we see an avoided crossing when L_{BB} is on resonance. If we plot the difference in the positions of these peaks vs the detuning of L_{BB} (Fig. 3.3d), we can see from the definitions of Δ_+ and Δ_- that the functional form will be:

$$\frac{1}{2}(\Delta_+ - \Delta_-) = \frac{1}{2}\sqrt{\delta_{2c}^2 + 4\Omega_{12}^2} \qquad (3.11)$$

By reading off the value of the above fitting function at $\delta_{2c} = 0$ we can therefore determine Ω_{12}, which will be important for determining transition strengths from photoassociation spectra (see Sect. 5.2.3). The agreement of the value Ω_{12} (which is a free fitting parameter in Eq. (3.10)) with the minimum of Eq. (3.11) serves as a reassuring consistency check for the validity of our fitting function.

3.3 Photo*dissociation* (Bound-Free) Spectroscopy

As is clear from Figs. 3.3 and 3.2, the lineshape of a photoassociation spectrum will be quite broad due to the spread collision energies with different thermal weights. We attempt to fully model the shape of this broadening with Eq. (3.7), but in doing so we make two critical assumptions:

1. The collision energies are distributed according to the Maxwell-Boltzmann distribution (Eq. (3.5)).
2. The atomic density is constant across the entire sample (Eq. (3.2)).

Each of the above assumptions stands on somewhat shaky ground.

With respect to the first, we know from experiments with thermometry in an optical lattice that Maxwell-Boltzmann statistics are a fairly good description for the initial distribution of energies in our trap [15], though for experiments operating nearer to quantum degeneracy a different model would have to be used. But our thermal sum over collision energies does not allow for the possibility of dynamical effects like frequency-dependent heating of our atoms by the photoassociation laser. In fact, strong heating of an atom cloud from a resonant photoassociation beam has already been reported in ultracold helium [11]. Our own measurements confirm that this effect is present in our experiment as well: see Fig. 3.4 for details.

The second assumption, i.e. that the atomic density is everywhere constant, is potentially even more worrisome. We interrogate atoms trapped in an optical lattice whose depth changes dramatically across its extent, which necessarily means that the density in different parts of the cloud must be different. Though shifts due to the trapping potential can in theory be modeled out [13], realistically this process

3.3 Photo*dissociation* (Bound-Free) Spectroscopy

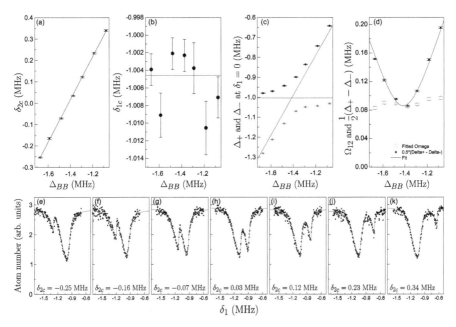

Fig. 3.3 Two-photon PA spectra are fitted with the combination of Eqs. (3.10) and (3.6). In order to cut down on computer processing time, Eq. (3.7) was not used (i.e., the trap was assumed to be exactly 2D). Panels (**a**)–(**d**) are plotted against Δ_{BB}, which is the experimentally controlled frequency offset of the bound-bound laser (L_{BB}) from an arbitrary reference point (ultimately determined by the length of our high-finesse cavity), and which is set by a programmable RF frequency driving an acousto-optic modulator (AOM). (**a**) Fitted value of the offset from resonance δ_{2c} of the bound-bound laser L_{BB} vs Δ_{BB}. (**b**) Fitted value of the offset from resonance δ_{1c} of the free-bound laser L_{FB} vs Δ_{FB}. (**c**) Locations of the left and right photoassociation peaks, given by the value of $\Delta_\pm$ (defined in the text) when $\delta_1 = 0$, vs Δ_{BB}. (**d**) $\frac{1}{2}(\Delta_+ - \Delta_-)$ vs Δ_{BB}. Traces (**e**)–(**k**) were taken by first changing the offset of the L_{BB} laser Δ_{BB}, and then sweeping the L_{FB} laser across the two-photon resonance. The data shown represents spectroscopy of the $X(-3, 0) \leftrightarrow 0_u(-6, 1)$ ($E = \sim 8430$ MHz) transition, and was recorded on September 23, 2014

will be imperfect, and can cause systematic uncertainties in the fit-determined value for the ground-state binding energy which are a significant fraction of the thermal linewidth.

The process of *photodissociation* on the other hand, describing the conversion of a bound molecule into a pair of atoms upon the absorption of a photon, does not suffer from either of these problems. The process is analogous to "photoassociation in reverse," with the primary difference being that there is no thermal weighting of the collision energies. Photodissociation is commonly used in experiments involving molecules as part of an imaging scheme, since it is usually far easier to photograph atoms (which possess strong cycling transitions) than molecules (which generally do not).

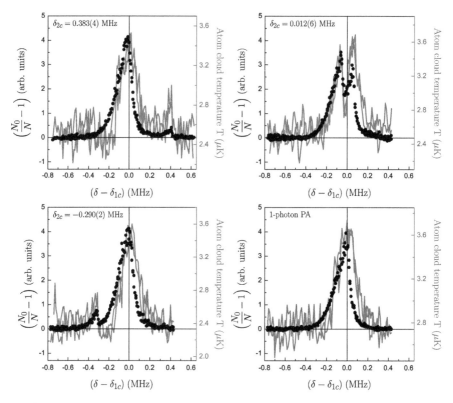

Fig. 3.4 Photoassociation spectra are recorded by first sweeping a ∼200 μW probe laser of duration 10 ms across resonance while atoms are trapped in a 1D optical lattice, then shutting off the lattice trapping beam and thereby releasing the atom cloud, and finally imaging the atom cloud after a ∼5.6 ms time of flight. Black points are measured values for $\left(\frac{N_0}{N} - 1\right)$ (where N_0 is the off-resonant atom number and N is the measured atom number at detuning $(\delta - \delta_{1c})$), which is proportional to the photoassociation rate (see Eq. (3.4)). Red lines are measured values of the temperature of the atom cloud remaining after photoassociation, as determined by a Gaussian fit to the cloud's spatial profile along the vertical axis assuming a Maxwell-Boltzmann distribution of velocities [1, 28]. Three of the panels show data for two-photon photoassociation spectra of the $X(-3, 0) \leftrightarrow 0_u(-4, 1)$ transition, with detuning of the bound-bound laser noted at upper left, while the final panel shows data for one-photon photoassociation of the $0_u(-4, 1)$ state. The discrepancy in off-resonant temperatures among graphs is likely due to a drifting magnetic field leading to a changing alignment between the MOT-cooled atom cloud and the optical lattice trap

In the following sections I will describe how we use photodissociation to perform accurate spectroscopy of both ground and excited state levels, with an absolute precision that can improve upon that achievable with photoassociation by a factor of 40 or more. For this thesis I will group photodissociation into two categories: "indirect," involving a transition to an electronic excited state which quickly decays

3.3 Photo*dissociation* (Bound-Free) Spectroscopy

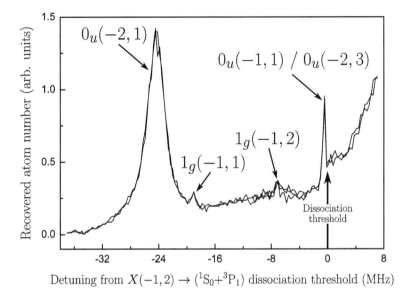

Fig. 3.5 Here is shown a spectrum resulting from irradiating $X(-1, 2)$ molecules with a $\sim$900 μW laser for 20 μs, recorded on January 13, 2015. Frequencies are defined with respect to the $X(-1, 2) \to (^1S_0 + {}^3P_1)$ dissociation threshold. The bumps in the spectrum are enhancements in the dissociation probability due to resonances with various bound states in the $^1S_0 + {}^3P_1$ manifold (labeled in the figure). The steep rise for $f > 0$ is representative of our empirical observation that the direct dissociation process is more efficient at larger frequencies (tens of MHz) above threshold than for smaller frequencies

to free atoms (i.e. the interrogation of *recovery* transitions); and "direct," involving the direct transition of a bound molecule into a pair of energetic atoms upon absorption of a photon. Figure 3.5 shows a particularly nice example of a spectrum showing both processes at work for a variety of different transitions.

3.3.1 Observing Photodissociation Gains

Whereas a photo*association* spectrum records *losses* from an atom cloud, a photo*dissociation* spectrum will show *gains* in atoms observed. To determine the functional form of the gain versus time, we first recognize that unlike photoassociation, the probability for a photon-molecule interaction to occur does not rely upon the atoms coming into close contact with one another, and therefore will not be proportional to n^2 (Eq. (3.2)). Instead, the rate will be proportional simply to N, the number of atoms in the cloud. We can then write the following differential equation describing photodissociation losses from the molecule cloud (ignoring long-timescale processes like "vacuum" losses):

$$\frac{dN(\delta)}{dt} = -K_{\text{eff}}^{\text{diss}}(\delta) \cdot N(\delta), \qquad (3.12)$$

which has the solution $N_\tau(\delta) = N_0 e^{-K_{\text{eff}}^{\text{diss}}(\delta)\tau}$. (Note that there is no difficulty here in converting from "density" to "number" since the differential equation is everywhere linear in N, guaranteeing that a photodissociation lineshape will not be negatively affected by making the questionable assumption of $n \propto N$.)

$N_\tau(\delta)$ describes the number of molecules remaining in the original cloud. However, what's actually observed is atoms resulting from the dissociation process. If we assume that our recovery process has an efficiency α, i.e. that only the fraction α of molecules which are dissociated end up as atoms which can be imaged, then the number of atoms observed will be:

$$N_\tau^{\text{obs}}(\delta) = \alpha N_0 (1 - e^{-K_{\text{eff}}^{\text{diss}}(\delta)\tau}) \qquad (3.13)$$

The lineshape, i.e. the dependence of N_τ^{obs} upon frequency, can then be determined from the frequency dependence of $K_{\text{eff}}^{\text{diss}}(\delta)$, which will have different forms depending on whether atoms are being gained via "recovery transitions" or by direct single-photon dissociation.

3.3.2 Recovery Transitions

For certain levels in Sr_2, with either conveniently placed binding energies or large transition moments to the free atom decay channel, a simplified scheme making use of only two lasers can be adopted for spectroscopy. The first creates a sample of ultracold molecules via photoassociation and subsequent spontaneous decay to a stable ground state molecule. The second excites a transition from the stable electronic ground state to a weakly bound excited state, and which is quickly converted to free atoms either by light-assisted photodissociation or spontaneous decay.

3.3.2.1 Spontaneous Decay

To determine the branching ratio governing what fraction of a molecular sample in some unstable electronic excited state $|\Psi_i\rangle$ will decay into a particular channel $|\Psi_f\rangle$, it is necessary to calculate the transition dipole moment $\langle\Psi_f|\vec{d}|\Psi_i\rangle$ and compare the amplitude squared of this moment with the amplitudes of transition moments to all other possible decay channels. For most bound states, spontaneous decay will be dominated by transitions to other bound states in the electronic ground state. However, if the initial state is very weakly bound, there can be a significant transition rate to free atoms.

In ^{88}Sr$_2$ it turns out that there are several levels with a significant branching ratio for decay to free atoms. We took advantage of this fact to make highly precise determinations of the Zeeman shifts of some of the most weakly bound levels in ^{88}Sr$_2$ in 2013 [17], and have since characterized their binding energies to high precision (see Table 3.1).

3.3.2.2 Two-Photon Dissociation

Spontaneous decay isn't the only pathway toward free atoms which can be induced with a single spectroscopy laser. If the binding energy of the final state is less than half the binding energy of the initial state, i.e. if $E_f \leq \frac{1}{2}E_i$, then the laser driving the transition from initial to final state can drive the final state to photodissociate into a pair of excited $^3P_1$ atoms, which will quickly decay into $^1S_0$ atoms which can subsequently be imaged in the normal way. This condition is illustrated schematically in Fig. 3.6.

If the only experimentally accessible variable when recording a spectrum were "atom number after recovery," then it would be difficult to discern how much of a role each of these two processes, spontaneous decay vs two-photon dissociation, plays in the total recovery signal. However, we have access to additional information: the spatial distribution of the atoms being recovered. Atoms produced via two-photon dissociation will have a well-defined kinetic energy, and therefore should form a ring expanding outward from the point of dissociation. Atoms produced via spontaneous decay should instead be emitted with a larger spread of energies, the exact details of which could be calculated by calculating the transition dipole moment's amplitude squared $|\langle \Psi_{\text{bound}}|\vec{d}|\Psi_{\text{free}}(\epsilon)\rangle|^2$, where the wavefunction describing free atoms $|\Psi_{\text{free}}(\epsilon)\rangle$ is a function of the kinetic energy of the fragments ϵ. An example of this difference in patterns is shown in the right-hand side of Fig. 3.6, which shows the recovery of $X(-1, 0)$ molecules via the 24 MHz $0_u(-2, 1, 0)$ state.

In either case, the dissociation rate K_{diss} should have the form of a Lorentzian, since it represents simply the probability of driving a transition between two bound states.

3.3.3 Direct One-Photon Photodissociation

Another option for recovering ground state molecules is their direct photodissociation into pairs of $^1S_0 + ^3P_1$ atoms. An accurate understanding of this process can also aid in accurately determining binding energies. In order to assign to vibrational levels *absolute* binding energies, we must have an absolute frequency reference against which they can be compared. The $^1S_0 + ^3P_1$ dissociation threshold is a

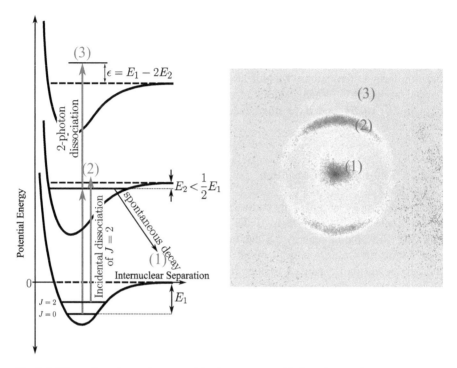

Fig. 3.6 When a laser is tuned to a transition between a weakly bound ground state of energy E_1 and a weakly bound excited state of energy $E_2 < \frac{1}{2}E_1$, the molecule can break apart into atoms either through spontaneous decay directly to the ground state or light-assisted photodissociation to the $^3P_1 + ^3P_1$ continuum. The left half of this figure shows schematically the possible routes toward dissociation, while the right half is an experimental image showing the resulting pattern of photofragments formed when a laser is tuned to the $X(-1,0)$ ($E = 136.6$ MHz) $\rightarrow 0_u(-2,1)$ ($E = 24.0$ MHz) transition. The dark middle ring labeled "(2)" is incidental dissociation of the $X(-1,2)$ ($E = 66.6$ MHz) state, present because our initial sample contains both $J = 0$ and $J = 2$ molecules

convenient choice, for if a bound state resonance occurs when a spectroscopy laser is tuned to f_{bound}, and the dissociation threshold occurs at $f_{^1S_0+^3P_1}$, then the binding energy of the bound state under investigation with respect to this threshold will be:

$$E_{\text{binding}} = -h \cdot (f_{^1S_0+^3P_1} - f_{\text{bound}}) \tag{3.14}$$

It is therefore of great interest to be able to spectroscopically determine the location of the dissociation threshold to high accuracy.

3.3.3.1 Derivation of "Shelf" Lineshape

In order to calculate a lineshape describing the dissociation threshold, we can appeal for inspiration to the single-photon photoassociation lineshape described by Eqs. (3.6) and (3.8), since single-photon association appears similar in many ways to single-photon dissociation with time running backwards. There are, however, some specific differences which must be considered.

One is that while photoassociation requires performing an integral over collision energies which are distributed according to the Boltzmann distribution, photodissociation requires no such thermal weighting of the output channel. This dramatically decreases the "linewidth" of the transition, with a concomitant improvement in the precision with which the transition can be inferred. We can incorporate this fact by setting the factor of $e^{-\frac{\epsilon}{kT}}$ in Eq. (3.5) equal to 1. We can then proceed to calculate the lineshape, which will have different forms depending on whether our trap geometry is two-dimensional or three-dimensional.

2D Lineshape Making the above substitution, using $g_{2D}(\epsilon)$ (which should be approximately valid for energies above threshold smaller than the axial trap spacing), and using $K_\epsilon \equiv K_{1\text{-photon}}$ (Eq. (3.8)) yields an equation which can be solved exactly [19]:

$$K_{\text{diss}}^{\text{eff, 2D}}(\delta_1) = C_{2D} \cdot \left\{ \frac{\pi}{2} + \tan^{-1}\left(\frac{2(\delta_1 - \delta_{1c})}{\gamma}\right)\right\}, \quad (3.15)$$

where C_{2D} is an overall scaling parameter and γ is an empirically determined linewidth.

3D Lineshape Equation (3.15) should be valid for small frequencies above threshold, where "reverse-collisions" occur at energies which are much smaller than the axial trap spacing. At very large energies above threshold, we can assume that the direction of the fragments is uninhibited by the dimensionality of the ~ 1 MHz deep lattice trap. Therefore in this case we must use $g_{3D}(\epsilon)d\epsilon = \frac{1}{2\pi^2}(\frac{2m}{\hbar^2})^{\frac{3}{2}}\epsilon^{\frac{1}{2}}d\epsilon$. Surprisingly, this scenario *also* admits an analytical solution:

$$K_{\text{eff,3D}} = C_{3D} \cdot \sqrt{(\delta_1 - \delta_{1c}) + \sqrt{(\delta_1 - \delta_{1c})^2 + (\gamma/2)^2}}. \quad (3.16)$$

Figure 3.7 shows an example of 1-photon "shelf recovery," in this case coming from data set used to calibrate the binding energy of the $1_g(-1, 1)$ state. Both the 2D "arctan" and 3D "square root" fits are shown, demonstrating that at least for small energies above threshold the 2D fit is a much better description of the data.

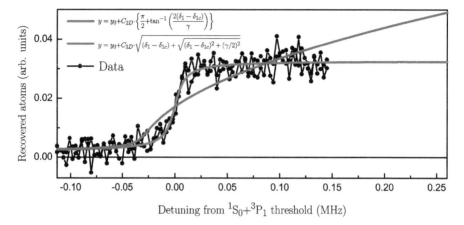

Fig. 3.7 $X(-1, 0)$ molecules are photodissociated via the $^1S_0+{}^3P_1$ threshold. The resulting spectrum is fit with either a 2D (red) or 3D (blue) function described in the text. For the 3D function, the linewidth γ was forced to equal the fitted value of the 2D fit (14.6 kHz) in order to aid in visual comparison. This trace is one of several used to calibrate the binding energy of the $1_g(-1, 1)$ state, and was recorded on April 24, 2014

3.3.3.2 Caveats and Assumptions

The derivations of Eqs. (3.15) and (3.16) make several simplifying assumptions which must be considered before absolute binding energies obtained via these equations can be absolutely trusted.

(1) Energy Dependence of $\gamma_s(\epsilon)$ In our quest to obtain a simple analytical fitting function, we assumed that the bound-free transition moment $\gamma_s(\epsilon)$ in Eq. (3.8) was constant across all relevant collision energies. However, we know that at large frequencies (i.e., 10s of MHz), the bound-to-free transition dipole moment $|\langle\Psi_i|\vec{d}|\Psi(\epsilon)\rangle|$ can vary dramatically due to the presence of shape resonances [16] and varying Franck-Condon factors. This problem can be minimized by focusing on very small detunings above threshold, where the transition moment would not be expected to vary by much.

(2) Recovery Efficiency of Dissociated Atoms At smaller frequencies, a more pressing concern lies in the ballistics of atoms escaping from the trap before we've had a chance to image them. In our experiments, we record the spectrum of a shelf by applying a long (several ms) dissociation pulse, and then after waiting for a few ms, image the remaining atom fragments in the usual way. At very large frequencies above the dissociation threshold, the atoms will possess enough kinetic energy to escape the ~1 MHz deep trap. At very small frequencies, nearly all should be captured. And at frequencies in between, the fraction remaining can be estimated by assuming a Boltzmann distribution for the energies of the initial molecules and carefully considering how the energy of fragments in the "lab frame" depends upon both the direction of the molecule and the direction of the emitted photofragments.

3.3 Photo*dissociation* (Bound-Free) Spectroscopy

Single-photon dissociation threshold at various lattice powers
(1 ms dissociation pulse followed by a 20 ms wait)

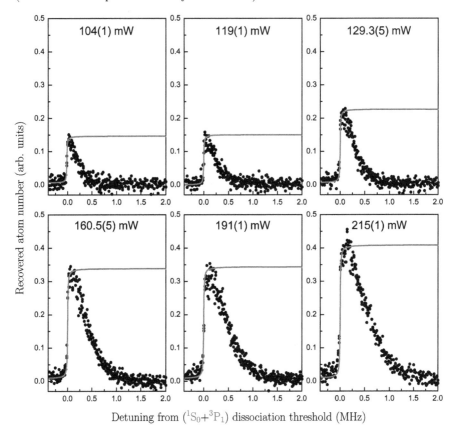

Fig. 3.8 Spectra of the one-photon dissociation threshold ($^1S_0+^3P_1$) are shown at various lattice powers. As lattice power (and therefore trap depth) increases, more atoms are observed at higher energies above threshold, since they are less likely to escape a deeper trap. The red curves are plots of Eq. (3.15), which describes the likelihood of a molecule photodissociating in 2D, but not the likelihood of an energetic atom being observed. In the future, spectra such as this may be used as a form of thermometry, since the steepness of fall-off of the "shelf tail" is related to the temperature of the molecules being dissociated

Figure 3.8 shows spectra describing the dissociation of molecules trapped in a 1D optical lattice at various lattice powers. I leave the derivation of this dependence as an exercise for the future grad student reader.

(3) Systematic Shifts Due to Inhomogeneous Lineshape Blurring The assumption that the photodissociation probability is given by a Lorentzian (i.e., Eq. (3.9)) ignores any possibility of inhomogeneous lineshape broadening. One source of broadening about which we're already aware (see Chap. 7) is due to *inhomogeneous lattice light shifts*, caused whenever the trap depths for initial versus final states are unequal. For transitions to electronic excited states this broadening should be minimal, since we generally operate at the magic wavelength for these transitions.

For two photon transitions between ground state levels, however, we will see that the lattice is certainly *not* magic, and that lattice-broadened lineshapes can have widths as large as $\sim$3 kHz. Additionally, it's also true that the lineshapes describing these two-photon dissociation processes are in fact much more complicated than those describing one-photon dissociation [29], with "linewidths" being determined by a combination of both lasers' powers and detunings.

For all shelf lineshapes used for the purposes of calibration to the dissociation threshold in this thesis, I have truncated the spectrum after a few tens of kHz in order to limit the possibility of systematic shifts due to recovery efficiency or the presence of shape resonances. Numerical simulations have shown that at most this procedure may result in shifts of $\sim$1 kHz. However, note that these precautions do not guard against the possibility of inhomogeneous lineshape broadening inducing systematic shifts, and therefore the interpretation of shelf lineshapes for the determination of ground state energies (described below) is potentially fraught with uncertainty.

3.3.4 Direct Two-Photon Photodissociation

Rather than inferring the binding energy of a ground state molecule via two-photon *association*, a potentially more precise method is to use two-photon *dissociation*, whereby a molecule is dissociated into two $^1S_0$ atoms by two lasers whose frequency difference is equal to the molecule's binding energy. See Fig. 3.1d for a schematic representation of this process. Just as in the case of one-photon dissociation, the "linewidth" of the transition will not be governed by an integral over thermally distributed collision energies. Instead, the lineshape will be determined by Eq. (3.5), with $K \equiv K_{\text{2-photon}}$ and the Boltzmann factor $e^{-\frac{\epsilon}{kT}}$ set equal to 1.

Whereas when dissociating to the $^1S_0+^3P_1$ threshold the linewidth is determined by the lifetime of the atomic excited state, in two-photon dissociation the final state consists of ground state atoms which are infinitely long-lived. While in principle this means that the dissociation linewidth can be made arbitrarily narrow, we are currently limited to $\sim$3 kHz widths due to ill-understood power broadening from our dissociation lasers and inhomogeneous broadening from a slightly non-magic lattice. See Fig. 3.9 for an example of the lineshape observed with this technique.

3.3.4.1 Comment on Lineshapes

In principle, lineshapes describing two-photon dissociation can be calculated analogously to how we proceeded for one-photon dissociation, i.e. by setting the Boltzmann weight $e^{\frac{\epsilon}{kT}}$ equal to 1 in Eq. (3.5) and integrating over all energies. Unfortunately, the integral of Eq. (3.10) with respect to ϵ doesn't admit a simple analytical solution, which meant that in order to fit photoassociation spectra (e.g.,

3.3 Photo*dissociation* (Bound-Free) Spectroscopy

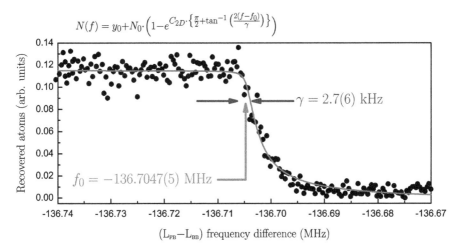

Fig. 3.9 A typical two-photon dissociation lineshape, fitted with Eqs. (3.13) and (3.15). This particular trace is one of many used to precisely determine the absolute binding energy of the $X(-1, 0)$ state (see Fig. 3.12), and was recorded on March 19, 2015 with the bound-bound laser detuned $\sim$20 to the red of the $X(-1, 0) \to 0_u(-4, 1)$ transition. Note that the width of this "shelf" lineshape is much narrower than that of the one-photon dissociation data shown in Fig. 3.7

in Fig. 3.3) we needed to perform a numerical integration, which is time-intensive (several minutes per fit) and somewhat unwieldy. But the resulting fit functions *did* rather beautifully describe our data, and we might ask whether it's worth pursuing a similar course for our photo*dissociation* spectra.

A key difference between these two experiments which makes the analogy imperfect is that while photoassociation spectroscopy is sensitive to atom *losses*, photodissociation spectroscopy is sensitive to atom *gains*. This means that with photodissociation we care not only about the interaction between laser and molecule describing the likelihood of a molecule producing fragments (the physics of which is described by Eq. (3.10)), but also about the dynamics of the fragments after dissociation, which will affect our probability of observing them.

We can reduce the importance of dynamical effects influencing our results by focusing only on energies close to threshold, where any photofragments produced are slow enough to be captured and imaged with near perfect fidelity. In our case that means probing energies above threshold of only a few tens of kHz in a lattice roughly 1 MHz deep. We also can arrange our spectroscopy lasers so that the fitting function should approximate the arctan fit given by Eq. (3.15). This should work well so long as the free-bound laser L_{FB} is far from resonance with the one-photon PA transition, and the difference between the frequencies of L_{FB} and L_{BB} is very close to the binding energy of the initial state. Put more concisely, if

$$(\Delta_1 - \Delta_2)^2 \gg \Omega_{12}^2, \tag{3.17}$$

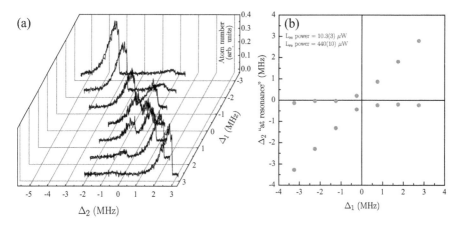

Fig. 3.10 (**a**) Spectra showing two-photon dissociation of the $X(-1, 0)$ state are shown, with the bound-bound laser at several small detunings with respect to the $0_u(-4, 1)$ state. Conventions for Δ_1 and Δ_2 are defined in Fig. 3.11. (**b**) Locations of the onset of photodissociation (i.e., rightmost edge of the spectroscopic "shelf" feature, labelled on the y-axis as "Δ_2 'at resonance' ") vs detuning of the bound-bound laser. For all data, the 2 ms photodissociation pulse was followed by a 20 ms wait, so the spectra reflect both the probability of photodissociation occurring and the likelihood that the photofragments will remain in the trap to be imaged

then the two-photon PA rate given by Eq. (3.10) reduces to the simple one-photon rate, whose integral yields the arctan fit given by Eq. (3.15). For the $X(-1, 0)$ "case study" described below, we have in fact used the arctan fit for all data analysis, which is somewhat justified because we have chosen Δ_1 to be very large ($\sim$20 MHz) and care only about fitting data very near threshold. However, it's likely that this assumption introduces small systematic errors into our final binding energy determination which will be discussed later.

3.3.4.2 Taming Highly Nonlinear Light Shifts

When $\Delta_1 \approx 0$, we would expect the two dissociation features to form an avoided crossing in analogy to Autler-Townes photoassociation spectroscopy. And indeed, when the bound-bound laser power is low, this is what we observe. Figure 3.10a shows several traces of the two-photon dissociation spectrum at various detunings of the bound-bound laser, and Fig. 3.10b plots the locations of the onset of these "shelves" vs laser frequency.

When the bound-bound laser power gets larger, however, strange things begin to happen. Figure 3.11 shows the locations of shelves versus bound-bound laser detuning for three different combinations of laser power. At large bound-bound laser powers, the Autler-Townes *doublet* apparently splits into a *triplet*, whose locations are highly nonlinear with laser frequency. Oddly, we have not observed similar

3.3 Photo*dissociation* (Bound-Free) Spectroscopy

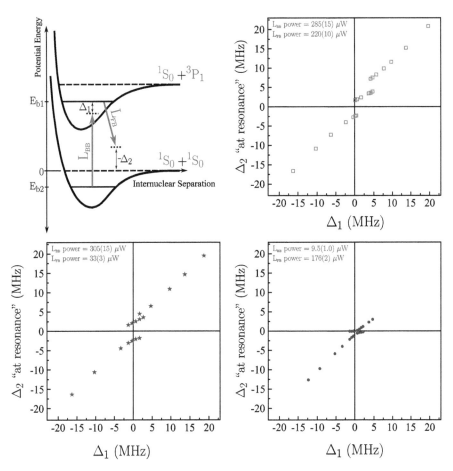

Fig. 3.11 The locations of two-photon dissociation "shelves" are plotted vs bound-bound laser detuning Δ_1 for a variety of laser powers. Note that the convention used to define Δ_1 and Δ_2 here differs from that used for two-photon photoassociation (described in Fig. 3.1b)

behavior with photo*association* spectroscopy, but this may be a result of the higher resolving power inherent to photodissociation absent the need to weight collision energies by a Boltzmann distribution.

The possible nonlinearity of light shifts near resonance is especially worrisome from the perspective of needing to make precision measurements extrapolating to zero power. In the case study described in the next section, we avoid this "danger region" by setting $\Delta_1 \approx -20$ MHz, far enough away to ensure that light shifts should be well-behaved.

3.3.4.3 Case Study: Binding Energy of $X(-1, 0)$

The reason we've been so careful in describing the possible pitfalls one might face in recording two-photon photodissociation spectra is because we'd like to use such spectra to make precise measurements of the absolute binding energies of ground state molecules. In particular, the most weakly bound rovibrational level of the electronic ground state is interesting because knowledge of its exact energy combined with the shape of the long-range interaction potential can be used to calculate the atomic scattering length [6, 8], which for ^{88}Sr is smaller than that of any of strontium's other stable isotopes.

To date, the most accurate measurements of ground-state binding energies have involved two-photon photo*association* of Sr atoms, but such measurements have only achieved precisions of a few hundred kHz [13, 24] due to reasons already discussed. These measurements, combined with theoretical calculations of the C_6, C_8, and C_{10} coefficients describing the long-range interaction potential, have constrained the scattering length of ^{88}Sr to be $a = -1.4(6)$ Bohr radii [13], where the uncertainty is in large part due to uncertainty in the absolute value of the binding energy.

In the following section I'll show that this precision can be improved by a factor of at least $\sim$40 with two-photon photo*dissociation* spectroscopy. To do so carefully requires evaluating several systematic effects which shift the position of the two-photon resonance, and then extrapolating these shifts to zero.

Magnetic Field Shifts Because the ground state of ^{88}Sr is nonmagnetic, stray magnetic fields of a few gauss will not shift the position of ground state rovibrational levels by an amount which can be detected in our experiment. A drifting magnetic field could, however, conceivably alter the light shifts induced by the spectroscopy lasers by shifting the $m = \pm 1$ sublevels of the intermediate state. In order to minimize this effect, a small (but constant) magnetic field of $\sim$1.1 gauss was applied during all measurements in order to define a stable quantization axis, and the detuning of the bound-bound laser was kept large so that the effect of magnetic shifts of far-away magnetic sublevels would be minimal.

Density Shifts Anecdotally, we have not yet observed in any of our measurements a clear dependence of frequency upon molecule density. This is in line with what we might expect for dilute molecules which carry no permanent dipole moment. We did try again here to observe a density shift by recording the threshold at "high" and "low" signal (accomplished by tuning the photoassociation laser used to create molecules away from perfect resonance), but found the slope of this dependence to be consistent with zero. Therefore in the calculations below, no corrections have been made for possible density shifts, since their magnitude is likely swamped by other sources of uncertainty.

RF Clock Uncertainty Our probe lasers are offset phase-locked to a stable master laser via RF synthesizers, which are themselves phase-locked to a stable

3.3 Photo*dissociation* (Bound-Free) Spectroscopy

10 MHz clock source. Since we define the binding energy of a rovibrational state operationally as the difference in laser frequencies required to drive transitions to an interesting bound state or the dissociation threshold, our value for the binding energy will depend critically upon how we define the RF frequency difference between these two lasers.

For all measurements described in this thesis, we used RF synthesizers locked to the 10 MHz clock output provided by an Stanford Research Systems SG384 equipped with a standard OXCO timebase. This synthesizer was calibrated in 2009, and is rated for less than 0.05 ppm aging per year. Measurements of the binding energy of the $X(-1, 0)$ state were made in March of 2015, implying (conservatively) an uncertainty in our absolute clock calibration of 6 years $\times$ 0.05 $\frac{\text{ppm}}{\text{year}}$ = 0.3 ppm. This translates to an absolute uncertainty of $\sim$0.3 ppm $\times$ 136.6 MHz $\approx$ 41 Hz, which is utterly negligible compared to the few-kHz uncertainties to be discussed in the next section. This RF calibration uncertainty *will*, however, be a dominant source of error when we evaluate binding energy differences among ground state levels (see Sect. 3.5).

Light Shifts Light shifts are the largest, most important systematic shift to evaluate. Determining the unperturbed energy then requires a triple extrapolation to zero power, because the molecules interact with three separate lasers: two for photodissociation and one for trapping.

Figure 3.12 shows how the position of the dissociation threshold depends on different laser powers. There are two halves to this figure (labelled "a" and "b") because this set of measurements was performed twice with very different laser frequencies. In panel (a), the bound-bound laser was tuned $\sim$20 MHz to the red of the $X(-1, 0) \rightarrow 0_u(-3, 1)$ transition, while in panel (b) it was tuned $\sim$20 MHz to the red of the $X(-1, 0) \rightarrow 0_u(-4, 1)$ transition. This guaranteed that the magnitudes (and in some cases even the signs) of light shifts induced by the three lasers for each of these experiments were substantially different. Then, if after independent extrapolations for each set of experimental parameters the final calculated binding energies disagree, we can attribute the disagreement to undiagnosed systematic effects.

For each row in the top half of this figure, one laser power was varied while the others were kept constant. In addition to the fitted "shelf" positions vs laser power, the fitted values of full-width-half-max (FWHM) are shown as well. Linear broadening with laser power could be a possible sign of systematic shifts, since we make no effort to model inhomogeneous broadening in our lineshapes, but rather simply use the arctan fit given by Eqs. (3.13) and (3.15). However, such shifts are almost certainly smaller than the measured width of the shelves at low laser power (i.e., a few kHz).

The bottom half of the figure shows the calculated "zero-power" binding energies for the $X(-1, 0)$ state for each data set shown in the top half of the figure. The binding energies are calculated for each point as the difference in laser frequencies

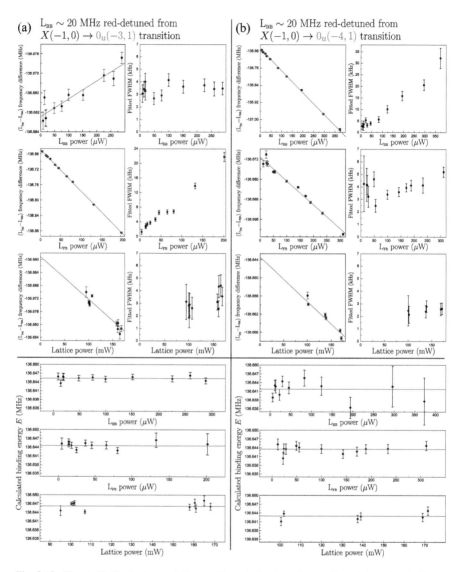

Fig. 3.12 (Top half) Evaluations of the position of the two-photon dissociation threshold are plotted vs bound-bound, free-bound, and lattice laser power. Measurements were performed twice, with the bound-bound laser tuned near the (**a**) $X(-1,0) \rightarrow 0_u(-3,1)$ or (**b**) $X(-1,0) \rightarrow 0_u(-4,1)$ transition. Red lines are linear fits to the data, and unfitted panels are plots of FWHM (as determined by a simple $\tan^{-1}$ fit) vs laser power. (Bottom half) Binding energies calculated from the above data by performing a triple extrapolation to zero power. The range of extrapolated binding energies from these six data sets spans less than 6 kHz

needed to drive the two-photon dissociation process minus the light shift due to each laser at that point. For all points shown in this figure, the laser powers were measured to <5% accuracy using a Thorlabs PM320E optical power meter with an S121B power head.

3.4 Bound-Bound Spectroscopy

Final Binding Energy Determination Performing a weighted average of the fitted binding energies shown in the bottom half of Fig. 3.12 gives the value $E = 136.64467$ MHz. This average clearly doesn't tell the whole story, since the measurements shown at left are uniformly a little higher than the measurements at right. However, their disagreement is small, and in fact is about the same size as the width γ of the dissociation threshold, which is about the size we might expect of systematic errors due to our imperfect fit function.

To accommodate both the left and right fitted values for the binding energy, we assign a conservative uncertainty of ± 5 kHz, yielding our final value for the binding energy of this state:

$$E_{X(-1,0)} = 136.6447(50) \text{ MHz} \cdot h. \tag{3.18}$$

This value represents an improvement by a factor of 40 over the current state of the art [13], and could be used to calculate more accurately the very small ^{88}Sr scattering length, possibly to an absolute precision competitive with that of the most accurate measurements for any atom [21].

3.4 Bound-Bound Spectroscopy

Bound-bound spectroscopy, involving transitions between rovibrational levels in the electronic ground and electronic excited state, represents an even more precise method for characterizing the locations of resonances, since (1) no integral over collision energy needs to be performed and (2) the lineshapes involved are symmetric. In order to determine the *absolute* binding energy of a state, however, it is necessary to compare the location of the (narrow) transition resonance described by a symmetric Lorentzian lineshape to the location of the (broad) dissociation threshold described by an asymmetric "shelf" lineshape. The measured binding energy will be affected by many systematic effects shifting the locations of both of these features, including shifts due to the optical lattice, the probe laser, magnetic fields, and molecule cloud density. To describe how these effects are evaluated, it's best to offer a case study for consideration.

3.4.1 Case Study: $1_g(-v, J' = 1)$

The most weakly bound state in the 1_g potential is extremely long-lived, with a lifetime of roughly 5 ms (see Chap. 6). This long lifetime implies that transitions to this state should correspondingly be extremely narrow, and makes it attractive as a "proving ground" for honing our abilities to do precision spectroscopy. As a case study, I will discuss the techniques we used to determine the binding energy of this state with an uncertainty of <5 kHz. This number is primarily limited by the uncertainty in determining the onset of the (~10 kHz wide) "shelf transition," though future experiments focusing instead on measuring the relative difference between two narrow subradiant states could be made even more accurately.

3.4.1.1 Correcting "Cavity Drift"

A frustrating technical issue in our experiment is that our main cavity-stabilized laser is not locked to an absolute external frequency reference, but instead is allowed to drift with the cavity. While the cavity was built to minimize these effects through vacuum shielding and active temperature feedback, we are still left with a residual cavity drift of several hundred hertz per minute. Since evaluating the strength of a systematic effect typically requires several tens of minutes to check the shift of a transition as an experimental parameter is varied, the cavity drift must be measured and corrected for.

To correct for this drift, data sets evaluating systematic shifts are interspersed with "calibration measurements," taken at a repeatable set of experimental conditions (i.e., the same lattice power, probe powers, magnetic field, etc.). Using timestamps from the logged data, we then plot the locations of the calibration measurements versus time and fit with a low-order polynomial (usually linear, but sometimes quadratic or higher). The polynomial fit is then used to subtract off a correction for every time-stamped measurement in the entire data set, ensuring that the calibration measurements do not move with time and that the shifts of all other data points are due only to the systematic effect being studied. For extremely precise lineshape studies, the frequency axis of individual traces is also "squeezed" in an analogous manner (see Fig. 7.6).

Note that for two-photon measurements sensitive only to frequency differences, cavity drift is unimportant. This is because when both lasers are locked to the same cavity (as is true when one is optically phase-locked to the other), drifts are common mode, and therefore cancel when evaluating their difference.

3.4.1.2 Zeroing the Magnetic Field

Rather than performing measurements with the magnetic field zeroed, we instead perform all binding energy measurements in the presence of an applied magnetic field which is small but large enough to properly set the quantization axis. We do this because there are two effects which compete in determining the quantization axis for our molecules: Zeeman shifts induced by the magnetic field, and tensor light shifts induced by our linearly polarized optical lattice. The quantization axis will be determined by whichever effect produces a larger energy splitting among different magnetic sublevels. See Fig. 3.13 for details.

In order to avoid nonlinear complications due to a changing quantization axis, we performed all measurements at an applied magnetic field of 430(30) mG. The uncertainty in this value comes from the procedure we use to zero the magnetic

3.4 Bound-Bound Spectroscopy

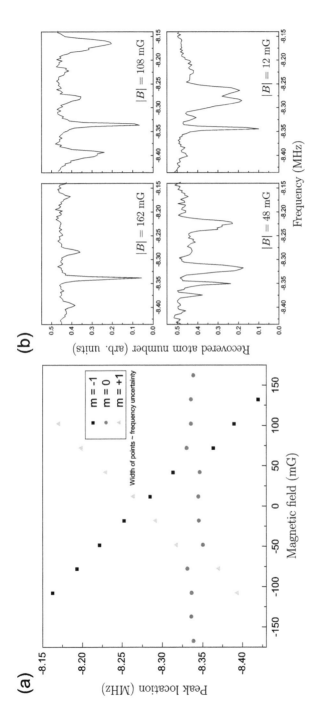

Fig. 3.13 (a) The locations of the $|m| = 0, 1$ sublevels of the $1_g(-1, 1)$ state are plotted against the applied magnetic field. Both the field direction and the lattice polarization direction are vertical. Note that at small magnetic fields there is an abrupt shift in the position of the $m = 0$ sublevel. We avoid this "danger region" by making all measurements of the binding energy for the $1_g(-1, 1)$ state at an applied magnetic field of 430 mG. This figure incorporates raw data described later in Fig. 4.5. (b) Spectra at selected magnetic fields, from which the data shown in (a) were taken. The prominent peaks are transitions to the $|m| = 0, 1$ sublevels, while the smaller bumps are sideband transitions

field, which is to find the value of applied magnetic field which makes the $m = \pm 1$ magnetic sublevels degenerate. At very small magnetic fields, the quantization axis is determined by the lattice polarization, and unfortunately in this case the trap is extremely non-magic for transitions to the $m = \pm 1$ sublevels. As a result, the lineshape describing transitions to these levels is very asymmetric and approximately 20 kHz broad. This limits the certainty with which we can say *definitively* whether these sublevels are absolutely energy-degenerate. Additionally, our magnetic field stabilization is very rudimentary, and prone to small drifts over the course of the day. This fact has been taken into consideration in assigning the magnetic field uncertainty above.

The frequency f of the transition to the $1_g(-1, 1, m = 0)$ state will shift quadratically with magnetic field according to $f = f_0 + \kappa B^2$. The uncertainty due to magnetic field can therefore be estimated as:

$$\Delta f \approx 2\kappa B \Delta B \tag{3.19}$$

The $1_g(-1, 1)$ is weakly bound, and therefore has a very large quadratic Zeeman shift of $\kappa = 121(3)$ kHz/G^2 (see Chap. 4). We can then estimate the uncertainty due to magnetic field as:

$$\Delta f \approx 2 \cdot 121 \, \frac{\text{kHz}}{\text{G}^2} \cdot 0.430 \, \text{G} \cdot 0.03 \, \text{G} = 3.1 \, \text{kHz} \tag{3.20}$$

Note that since the uncertainty is proportional to both B and κ, this uncertainty will be smaller for more deeply bound states with smaller quadratic Zeeman shifts, and can be minimized for any state by decreasing the B-field (so long as the quantization axis remains well-defined).

3.4.1.3 Nearly-Magic Lattice Light Shifts

Care was taken to arrange the optical lattice so that it was as nearly magic as possible for the transition under investigation. Unfortunately, the "magic lattice" condition for a transition to the $1_g(-1, 1, m = 0)$ state is different from that which is required for the $^1S_0 + ^3P_1$ dissociation threshold. A future experiment could improve precision by quickly rotating the lattice laser polarization between measurements of the $1_g(-1, 1, m = 0)$ state and the $^1S_0 + ^3P_1$ dissociation threshold so that it remained approximately magic in both cases. For the data presented here, however, the total shift was simply measured for each feature and calibrated out in order to determine the final binding energy.

Figure 3.14 shows the lattice light shift for both the $1_g(-1, 1, m = 0)$ state and the $^1S_0 + ^3P_1$ dissociation threshold. The magnitude of this shift was measured by recording the location of each feature at alternately high and low lattice power (i.e., trap depth), correcting the data for cavity drift, and then fitting a line to the resulting plot of frequency vs lattice power.

3.4 Bound-Bound Spectroscopy

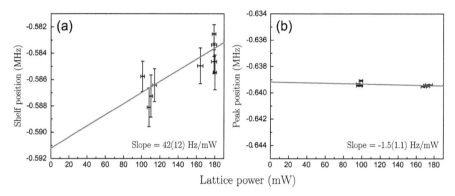

Fig. 3.14 Because the "magic lattice" conditions for the $X(-1, 0) \to 1_g(-1, 1, 0)$ and $^1S_0 \to {}^3P_1$ transitions are mutually exclusive (i.e., they require different lattice laser polarization orientations), we can only minimize lattice light shifts for one of these transitions. For the measurements described in this section, we arranged for the lattice to be magic for $X(-1, 0) \to 1_g(-1, 1, 0)$. (**a**) Location of the $^1S_0 \to {}^3P_1$ "shelf" transition (as determined by a simple $\tan^{-1}$ fit) plotted vs lattice power. (**b**) Location of the $X(-1, 0) \to 1_g(-1, 1, 0)$ transition (as determined by a Lorentzian fit) plotted vs lattice power. While the shift in (**b**) is consistent with zero, the shift in (**a**) is large enough to induce an error of $\sim$5 kHz if not properly accounted for

The resulting light shift for the $1_g(-1, 1, m = 0)$ state is very small ($-1.5(1.1)$ Hz/mW), resulting in an overall shift of $\sim$255(187) Hz compared to the zero lattice-power location. The shift for the $^1S_0 + {}^3P_1$ dissociation threshold is larger at 42.2(13.1) Hz/mW, resulting in a net shift of 7.6(2.2) kHz for our 180 mW lattice. It is clear that the dominant contribution to the uncertainty in this measurement comes from uncertainty in measurement of the lattice light shift of the $^1S_0 + {}^3P_1$ dissociation threshold.

3.4.1.4 Probe Light Shifts

The probe lasers can also induce (very small) light shifts in the features under examination. However, these light shifts can be made to be extremely small by using very small probe powers and simply increasing the interrogation time. (Note that this option is not available in the case of lattice power, since the minimum possible trap depth is set by the $\sim$3 μK temperature of our trapped molecules, and higher lattice powers generally lead to higher signal-to-noise—see Fig. 3.8).

Figure 3.15 shows plots of the light shifts measured for these states. Note that the light shifts were measured by increasing the probe power to much higher levels than was actually used for the final determination of binding energy, so that the actual shift is very small.

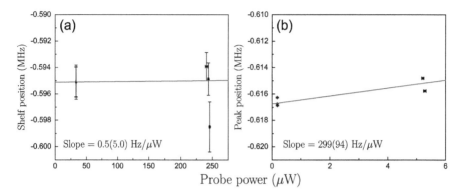

Fig. 3.15 (a) Location of the $^1S_0 \to {}^3P_1$ "shelf" transition (as determined by a simple $\tan^{-1}$ fit) plotted vs probe laser power. (b) Location of the $X(-1,0) \to 1_g(-1,1,0)$ transition (as determined by a Lorentzian fit) plotted vs probe laser power

3.4.1.5 RF Clock Uncertainty

Measurements of the binding energy of the $1_g(-1,1)$ state were made on April 24, 2014. As was discussed in Sect. 3.3.4.3, the uncertainty due to 0.05 ppm aging per year of our SG384's OXCO ("oven-controlled crystal oscillator") internal timebase last calibrated in 2009 will be of the order of only a few Hz, and therefore completely negligible for this particular measurement.

3.4.1.6 Final Calculation of Binding Energy

Extrapolating shifts for both the $1_g(-1,1)$ state and the $^1S_0 + {}^3P_1$ shelf to zero power and field, and then calculating the difference in laser frequencies between these two resonance conditions, yields the following value for the binding energy:

$$E_{1_g(-1,1)} = 19.0420(38)\,\text{MHz} \cdot h \quad (3.21)$$

The $\sim$4 kHz uncertainty in this value is dominated mostly by two systematics: the uncertainty in the absolute magnetic field (yielding $\sim$3.1 kHz), and the uncertainty in the lattice laser light shift of the dissociation threshold (yielding $\sim$2.2 kHz). The magnetic field uncertainty can in the future be minimized by either operating at a smaller magnetic field, or choosing to observe a state with a smaller quadratic Zeeman shift (or by discovering a better way to characterize and stabilize the magnetic field within the chamber). The uncertainty in the lattice light shift of the dissociation threshold is more difficult to contend with, but could be minimized by making measurements under magic lattice conditions (perhaps by rapidly switching lattice polarization between measurements of the positions of the shelf and the $1_g(-1,1)$ state).

3.5 Coherent Two-Photon Raman Transitions (Ground State Binding Energy Differences)

The most accurate measurements our lab can currently make (in terms of absolute frequency uncertainty) turn out to be relative differences in the binding energies of different rovibrational levels in the electronic ground state. This is because molecules in the electronic ground state are inherently stable, so that the linewidth of the transitions we can observe are not limited by finite lifetime of the final state. These measurements are made by determining the frequency difference between lasers required to drive a two-photon Raman transition between levels with binding energies E_{b1} and E_{b2} (see Fig. 3.16), extrapolating to the limit of zero power for all lasers.

The achievable precision with which we can measure frequency differences is ultimately limited by the linewidth we can achieve for the 2-photon Raman transition. This linewidth is related to the coherence time characterizing Rabi oscillations between the two levels under consideration.

3.5.1 Comments on Coherence Time

Figure 3.17 shows Rabi oscillations between the $X(-2, 0)$ and $X(-1, 0)$ levels under various conditions. To observe these oscillations, we (1) prepare molecules in a state with binding energy E_{b1}, (2) subject them to a 2-color probe pulse of duration τ and

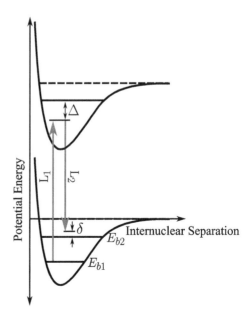

Fig. 3.16 A schematic representation of the scheme used to excite Raman transitions between bound states. For the measurements described in this section, Δ is several hundred MHz from the nearest bound state, and binding energy differences represent the extrapolated zero-power difference in frequencies between L_1 and L_2 when $\delta = 0$

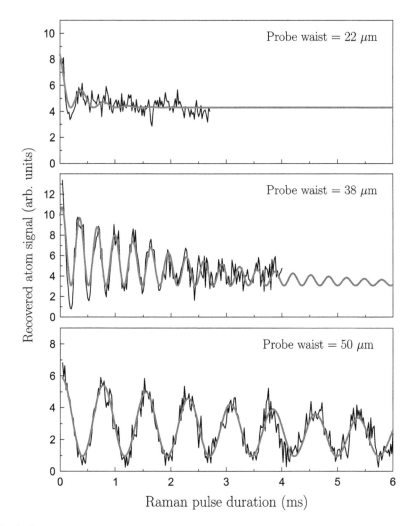

Fig. 3.17 The coherence time of a two-photon Raman transition is dramatically affected by the waist of the probe lasers used to drive it, since smaller waists can result in inhomogeneous laser intensity across the molecule cloud. Here are shown Rabi oscillations observed for the $X(-2, 0) \to X(-1, 0)$ transition for three different values of the probe waist

detuning $\delta = 0$ as depicted in Fig. 3.16, and (3) count the molecules remaining in the initial state by photodissociation and absorption imaging of the resulting atomic fragments. Figure 3.17 illustrates that the primary impediment to long coherence times is inhomogeneous probe laser intensity across the atom sample. This limitation can be minimized along the transverse direction by expanding the probe waist so that the entire molecule cloud sees a more homogeneous probe beam.

3.5 Coherent Two-Photon Raman Transitions (Ground State Binding Energy...

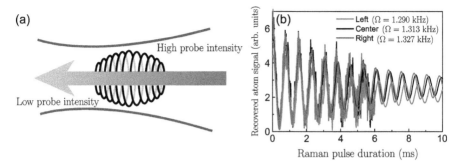

Fig. 3.18 (a) As the probe beam passes through the molecular cloud, it can become attenuated, causing molecules near the front of the cloud to cycle through Rabi oscillations more quickly than molecules near the back. (b) We observe hints of this effect by measuring the population of a particular electronic ground state vs probe duration at three different points in the cloud, as viewed from a camera oriented perpendicular to the lattice trapping and probe axis. The fitted Rabi frequency for molecules at the left of the cloud is observed to be slightly smaller than for molecules at the right

Another effect which is not so easy to mitigate is the fact that as the probe beam travels through the molecule cloud, it will be partially absorbed. Therefore molecules near the "front" of the cloud will see a more intense beam, while molecules near the "back" will see a less intense beam. Figure 3.18 illustrates the problem schematically, and shows some rough data confirming that the Rabi frequency can vary by approximately 3% across the cloud. The importance of this effect can be minimized, however, by simply operating at smaller probe powers (and therefore smaller Rabi frequencies), since the degree of decoherence will be smaller if the molecules oscillate through fewer Rabi cycles.

3.5.2 Determination of Binding Energy Differences Among $X(-1, 0)$, $X(-2, 0)$, and $X(-3, 0)$ States

We have so far precisely measured binding energy differences between two pairs of levels in the electronic ground state [19]:

- $E_{X(-1,0)} - E_{X(-2,0)} = 1263.673582 \pm (63)_{\text{exp}} \pm (320)_{\text{cal}}$ MHz
- $E_{X(-2,0)} - E_{X(-3,0)} = 3710.255610 \pm (170)_{\text{exp}} \pm (930)_{\text{cal}}$ MHz

Here, errors labeled "exp" result from uncertainty in extrapolating resonance frequencies to zero laser power, while errors labeled "cal" result from imperfect calibration of our SG384's OXCO internal clock. The following sections clarify how these systematic effects were evaluated.

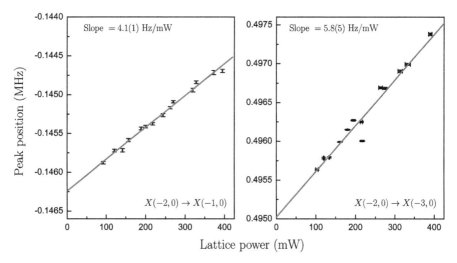

Fig. 3.19 The location of the two-photon resonance for $X(-2, 0) \rightarrow X(-1, 0)$ (left) and $X(-2, 0) \rightarrow X(-3, 0)$ (right) as defined by the detuning of one of the two spectroscopy lasers from an arbitrary reference point is plotted against lattice power

3.5.2.1 Dominant Sources of Uncertainty

Lattice Light Shifts Though the optical lattice is approximately magic for the $^1S_0 + {}^3P_1$ atomic transition, it is in general slightly non-magic for transitions between molecular states. When one of the states involved in the transition possesses a rotational momentum $J \neq 1$, it is often possible to balance tensor and scalar light shifts such that the net differential shift is zero (see "Subradiant Spectroscopy"). For two-photon transitions between $J = 0$ molecules in the electronic ground state, however, no tensor light shift is present, and therefore our options are more limited. It is believed that magic wavelengths *should* exist for pairs of rovibrational levels which would be interesting for the construction of a molecular clock [31]. The development of magic-wavelength traps for these molecular clock transitions will therefore be an important task for future experimental work. For this thesis, however, we simply measured the (small) differential lattice light shift and extrapolated to zero lattice power.

Figure 3.19 shows the shift in the two-photon resonance as a function of lattice power. Note that though the shift is extremely small (only 4.1(1) Hz/mW), we can nevertheless measure it with impressive precision. This is due in equal parts to the facts that the raw data used to construct this plot consisted of very narrow spectroscopic traces (linewidth ≈ 300 Hz), and that cavity drift does not influence differential frequency measurements.

Probe Light Shifts Because two lasers are simultaneously applied to the molecules, we must separately determine the shifts imparted by each, and then extrapolate to zero power. Figure 3.20 shows the shift of the resonance location

3.5 Coherent Two-Photon Raman Transitions (Ground State Binding Energy...

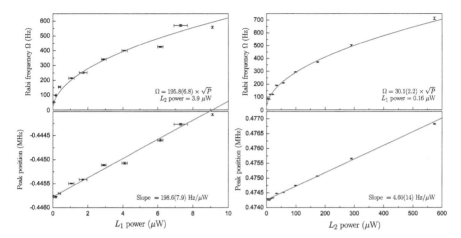

Fig. 3.20 The Rabi flopping rate for a two-photon transition between levels in the electronic ground state will be proportional to the square root of the product of laser powers used for spectroscopy. In the top row, this Rabi flopping rate is plotted against each laser power. In the bottom row, the location of the two photon resonance is plotted against each laser power. All data is for the $X(-2, 0) \to X(-1, 0)$ transition

vs power for both probe lasers L_1 and L_2. The total shift was assumed to simply be the sum of the shifts separately induced by each probe laser, which should be a safe assumption so long as L_1 is tuned far from resonance with the intermediate state (i.e., so long as all light shifts are linear), which was indeed the case for this experiment. In this case, the light shift due to L_1 can be measured by setting the power of L_2 to a small value, varying the power of L_1, and recording the position of the 2-photon resonance.

Inhomogeneous Lineshape Fitting These measurements were made in May, 2013, before we fully understood the nature of how the small differential lattice light shift acted upon our lineshape. We now know that for molecules at $\sim 3\,\mu\text{K}$ in an approximately harmonic optical lattice trap, the lineshape of the two-photon resonance will be asymmetric, with a linewidth approximately one third of the total shift induced by the lattice trap [15]. At the time this data was analyzed, we did not account for the asymmetric nature of this lineshape, but instead fit our data with a simple Lorentzian function. In cases where the molecule cloud temperature does not asymptotically approach zero at zero lattice power, we know that this can induce a systematic error in the determination of the unperturbed resonance location. In our experiment, however, the molecule temperature *does* closely approach zero at zero lattice power (see Fig. 7.7a). Therefore we don't expect the error due to an incorrect fitting function to be significant.

RF Electronics Calibration Measurements of both binding energy differences described at the beginning of this section were made within 1 month of one another in 2013. They therefore both suffer from approximately the same fractional error due to an aging RF clock, and because the frequencies involved are so large (and the other experimental uncertainties so small), RF clock calibration contributes to a very significant uncertainty in our reported binding energy. Conservatively assuming a fractional uncertainty for the >4 year drift of 0.25 ppm, we calculate an uncertainty of:

- $\Delta(E_{X(-1,0)} - E_{X(-2,0)})_{cal} \approx 1264\,\text{MHz} \times 0.25\,\text{ppm} \approx 320\,\text{Hz}$
- $\Delta(E_{X(-2,0)} - E_{X(-3,0)})_{cal} \approx 3710\,\text{MHz} \times 0.25\,\text{ppm} \approx 930\,\text{Hz}$

These uncertainties are larger than any other source of experimental uncertainty, and therefore dominate our error budget. But note that this is an easy problem to fix—future measurements can reduce this number by simply calibrating the RF electronics to a better clock. This is in fact already being done in our lab, as we now discipline our 10 MHz clock to GPS.

References

1. Arora, P., Purnapatra, S., Acharya, A., Kumar, R., Gupta, A.: Measurement of temperature of atomic cloud using time-of-flight technique. MAPAN **27**(1), 31–39 (2012)
2. Bohn, J., Julienne, P.: Semianalytic treatment of two-color photoassociation spectroscopy and control of cold atoms. Phys. Rev. A **54**(6), R4637 (1996)
3. Borkowski, M., Morzyński, P., Ciuryło, R., Julienne, P., Yan, M., DeSalvo, B., Killian, T.: Mass scaling and nonadiabatic effects in photoassociation spectroscopy of ultracold strontium atoms. Phys. Rev. A **90**(3), 032713 (2014)
4. Chin, C., Flambaum, V.: Enhanced sensitivity to fundamental constants in ultracold atomic and molecular systems near Feshbach resonances. Phys. Rev. Lett. **96**(23), 230801 (2006)
5. Cohen-Tannoudji, C.: The Autler-Townes effect revisited. In: Amazing Light, pp. 109–123. Springer, Berlin (1996)
6. Crubellier, A., Dulieu, O., Masnou-Seeuws, F., Elbs, M., Knöckel, H., Tiemann, E.: Simple determination of Na_2 scattering lengths using observed bound levels at the ground state asymptote. Eur. Phys. J. D **6**(2), 211–220 (1999)
7. Friebel, S., D'andrea, C., Walz, J., Weitz, M., Hänsch, T.: CO_2-laser optical lattice with cold rubidium atoms. Phys. Rev. A **57**(1), R20 (1998)
8. Gao, B.: Zero-energy bound or quasibound states and their implications for diatomic systems with an asymptotic van der Waals interaction. Phys. Rev. A **62**(5), 050702 (2000)
9. Jones, K., Tiesinga, E., Lett, P., Julienne, P.: Ultracold photoassociation spectroscopy: long-range molecules and atomic scattering. Rev. Mod. Phys. **78**(2), 483 (2006)
10. Kamiya, Y., Itagaki, K., Tani, M., Kim, G., Komamiya, S.: Constraints on new gravitylike forces in the nanometer range. Phys. Rev. Lett. **114**(16), 161101 (2015)
11. Léonard, J., Mosk, A., Walhout, M., van der Straten, P., Leduc, M., Cohen-Tannoudji, C.: Analysis of photoassociation spectra for giant helium dimers. Phys. Rev. A **69**(3), 032702 (2004)
12. Lutz, J., Hutson, J.: Deviations from Born-Oppenheimer mass scaling in spectroscopy and ultracold molecular physics. J. Mol. Spectrosc. **330**, 43–56 (2016)
13. Martinez de Escobar, Y., Mickelson, P., Pellegrini, P., Nagel, S., Traverso, A., Yan, M., Côté, R., Killian, T.: Two-photon photoassociative spectroscopy of ultracold ^{88}Sr. Phys. Rev. A **78**, 062708 (2008)
14. McDonald, M.: High precision optical spectroscopy and quantum state selected photodissociation of ultracold $^{88}Sr_2$ molecules in an optical lattice. Ph.D. Thesis, Columbia University in the City of New York (2016)

References

15. McDonald, M., McGuyer, B., Iwata, G., Zelevinsky, T.: Thermometry via light shifts in optical lattices. Phys. Rev. Lett. **114**(2), 023001 (2015)
16. McDonald, M., McGuyer, B., Apfelbeck, F., Lee, C.-H., Majewska, I., Moszynski, R., Zelevinsky, T.: Photodissociation of ultracold diatomic strontium molecules with quantum state control. Nature **534**(7610), 122–126 (2016)
17. McGuyer, B., Osborn, C., McDonald, M., Reinaudi, G., Skomorowski, W., Moszynski, R., Zelevinsky, T.: Nonadiabatic effects in ultracold molecules via anomalous linear and quadratic Zeeman shifts. Phys. Rev. Lett. **111**(24), 243003 (2013)
18. McGuyer, B., McDonald, M., Iwata, G., Skomorowski, W., Moszynski, R., Zelevinsky, T.: Control of optical transitions with magnetic fields in weakly bound molecules. Phys. Rev. Lett. **115**(5), 053001 (2015)
19. McGuyer, B., McDonald, M., Iwata, G., Tarallo, M., Grier, A., Apfelbeck, F., Zelevinsky, T.: High-precision spectroscopy of ultracold molecules in an optical lattice. New J. Phys. **17**(5), 055004 (2015)
20. McGuyer, B., McDonald, M., Iwata, G., Tarallo, M., Skomorowski, W., Moszynski, R., Zelevinsky, T.: Precise study of asymptotic physics with subradiant ultracold molecules. Nat. Phys. **11**(1), 32–36 (2015)
21. Moal, S., Portier, M., Kim, J., Dugué, J., Rapol, U., Leduc, M., Cohen-Tannoudji, C.: Accurate determination of the scattering length of metastable helium atoms using dark resonances between atoms and exotic molecules. Phys. Rev. Lett. **96**(2), 023203 (2006)
22. Napolitano, R., Weiner, J., Williams, C., Julienne, P.: Line shapes of high resolution photoassociation spectra of optically cooled atoms. Phys. Rev. Lett. **73**(10), 1352 (1994)
23. Osborn, C.: The physics of ultracold Sr_2 molecules: Optical production and precision measurement. Ph.D. Thesis (2014)
24. Reinaudi, G., Osborn, C., McDonald, M., Kotochigova, S., Zelevinsky, T.: Optical production of stable ultracold $^{88}Sr_2$ molecules. Phys. Rev. Lett. **109**, 115303 (2012)
25. Salumbides, E., Koelemeij, J., Komasa, J., Pachucki, K., Eikema, K., Ubachs, W.: Bounds on fifth forces from precision measurements on molecules. Phys. Rev. D **87**(11), 112008 (2013)
26. Skomorowski, W., Pawłowski, F., Koch, C., Moszynski, R.: Rovibrational dynamics of the strontium molecule in the $A^1\Sigma_u^+$, $c^3\Pi_u$, and $a^3\Sigma_u^+$ manifold from state-of-the-art ab initio calculations. J. Chem. Phys. **136**(19), 194306 (2012)
27. Tarallo, M., Iwata, G., Zelevinsky, T.: BaH molecular spectroscopy with relevance to laser cooling. Phys. Rev. A **93**(3), 032509 (2016)
28. Weiss, D., Riis, E., Shevy, Y., Ungar, P., Chu, S.: Optical molasses and multilevel atoms: experiment. J. Opt. Soc. Am. B **6**(11), 2072–2083 (1989)
29. Zanon-Willette, T., De Clercq, E., Arimondo, E.: Ultrahigh-resolution spectroscopy with atomic or molecular dark resonances: exact steady-state line shapes and asymptotic profiles in the adiabatic pulsed regime. Phys. Rev. A **84**(6), 062502 (2011)
30. Zelevinsky, T., Boyd, M., Ludlow, A., Ido, T., Ye, J., Ciuryło, R., Naidon, P., Julienne, P.: Narrow line photoassociation in an optical lattice. Phys. Rev. Lett. **96**(20), 203201 (2006)
31. Zelevinsky, T., Kotochigova, S., Ye, J.: Precision test of mass-ratio variations with lattice-confined ultracold molecules. Phys. Rev. Lett. **100**(4), 043201 (2008)

Chapter 4
Measurements of Zeeman Shifts

4.1 Introduction and Summary of Measurements

The term *Zeeman shift* refers to the perturbation of atomic or molecular energy levels due to the presence of a magnetic field, and was first observed by Pieter Zeeman in 1897 [21]. In those original observations, Zeeman noticed that certain spectral lines in sodium would become broader in the presence of a magnetic field. Even more dramatically, the light emitted from the edges of these lines was circularly polarized, hinting at a connection between angular momentum and the "vibrations" of the particles which emitted the light. These observations (as well as a partial explanation of their origins) earned Pieter Zeeman and Hendrik Lorentz a Nobel Prize in 1902. But whereas more than 100 years ago measurements were only able to resolve a slight blurring and change in the degree of polarization from magnetically-broadened lines, today we can do quite a bit better.

Considering that our experiment was designed to adapt techniques originally intended for the construction of the most accurate clocks in the world to the task of molecular spectroscopy, it's fair to say that we take pride in making extremely precise measurements. The best clocks in the world routinely interrogate spectral features at the part in 10^{18} level (a record which will almost certainly be broken soon after this thesis is published), and our own lab hopes to eventually measure binding energy differences to within a few orders of magnitude of this precision. To make measurements of molecular binding energies so accurately requires isolating molecules from external perturbations, so that the resulting measurements are sensitive only to universal molecular physics rather than the vagaries of the local environment. But an accurate understanding of how the environment affects molecular structure is of course a necessary step toward achieving this goal, and moreover can reveal aspects of molecular physics which are plenty interesting in their own right.

Molecular Zeeman shifts are one such "environmental effect" which can reveal a great deal about molecular structure. And whereas a single, highly precise measurement of an experimental quantity can reveal the answer to specific, targeted question, examining patterns in larger data sets (and when these patterns are broken) can reveal general trends and deeper rules. For this reason, I include in the introduction to this chapter four Tables (Tables 4.1, 4.2, 4.3, and 4.4) summarizing linear and quadratic Zeeman shifts for the majority of levels so far discovered. Two patterns are immediately apparent:

- **Cells highlighted in green have approximately the same linear Zeeman shift coefficients.** And whether the green cells belong to the 1_u or 1_g potentials depends on whether the rotational angular momentum J is *odd* or *even*.
- **Quadratic Zeeman shifts among levels with the same character (1_u, 0_u, 1_g) increase as states become more weakly-bound.** And a more careful look shows that the quadratic Zeeman shifts scale approximately with (bond length)$^{\sim 3}$.

These observations hint at something deeper than coincidence, and the following sections will aim to connect these trends to a deeper, more general understanding of molecular physics.

4.1.1 Details About Data Presented in Tables 4.1, 4.2, 4.3, and 4.4

Tables 4.1, 4.2, 4.3, and 4.4 present linear and quadratic Zeeman shifts for the majority of all singly-excited states of ^{88}Sr$_2$ with $J' \leq 4$ and $E_b < 8.5$ GHz. Green cells represent levels for which the ideal Hund's case (c) Zeeman shift calculation should apply exactly (explained in Sect. 4.2.3 below). The column heading "(B_i, B_f) (G)" refers to the magnetic field range over which Zeeman shift data was recorded.

For all entries in these tables, Zeeman coefficients were extracted from a fit of the peak positions versus magnetic field with the function

$$f(B) = f_0 + \sum_{i=1}^{i=6} \left(\frac{m}{|m|}\right)^i \cdot \beta_i \cdot B^i \tag{4.1}$$

where m is the magnetic sublevel being studied and the factor $\left(\frac{m}{|m|}\right)^i$ is to ensure the correct sign for positive and negative sublevels. For the case of $m = 0$, $\beta_{(\text{odd})}$ were forced to equal 0 because of symmetry.

For the large majority of cases, Zeeman shift coefficients were extracted from a fit of peak positions vs magnetic field which forced $(\beta_3, \beta_4, \beta_5, \beta_6) = 0$. Entries marked with "*", however, allowed higher fit coefficients. Whether higher

4.1 Introduction and Summary of Measurements

orders than quadratic should be used was determined by whether the residuals of the fit appeared to be randomly distributed about zero. For the full magnetic field dependence of these starred entries, see Table 4.5.

[a] For the quadratic Zeeman shift of the $m = 0$ sublevel of the $0_u(-1, 1)$ state, the value published in [10] does not properly account for the effect of tensor light shifts, which cause the transition frequency to artificially jump at very small values of magnetic field as the magnetic field no longer defines the quantization axis (see, e.g., Fig. 3.13). The value quoted in this table represents a new analysis of the same data, but properly accounts for this effect and is therefore more trustworthy.

Table 4.1 Linear and quadratic Zeeman shifts for all singly-excited rovibrational levels of $^{88}Sr_2$ with $E < 8.5$ GHz and $J' = 1$

$J' = 1$	E (MHz)	$\|m\|$	β_1 (MHz/G)	β_2 (MHz/G^2)	(B_i, B_f) (G)	Ref.
$0_u(-1, 1)$	0.4653(45)	0	X	−0.355(31)	(−0.3, +0.5)	McGuyer et al. [10][a]
		1	0.932(19)	−0.764(61)	(−0.4, +0.3)	McGuyer et al. [10]
$1_g(-1, 1)$	19.0420(38)	0	X	−0.121(3)	(−1.8, 2.0)	McGuyer et al. [12]
		1	1.048(2)	−0.065(2)	(−1.8, 2.0)	McGuyer et al. [12]
$0_u(-2, 1)$	23.9684(50)	0	X	−0.0253(8)	(−4.0, 1.0)	McGuyer et al. [10]
		1	0.3247(44)	−0.0621(18)	(−4.0, 1.0)	McGuyer et al. [10]
$0_u(-3, 1)$	222.161(35)	0	X	−0.00323(88)	(−4.0, 1.0)	McGuyer et al. [10]
		1	0.2248(32)	−0.00925(58)	(−4.0, 1.0)	McGuyer et al. [10]
$1_g(-2, 1)$	316(1)	0	X	−0.0269(4)	(−1.8, 2.0)	McGuyer et al. [12]
		1	1.042(3)	−0.019(3)	(−1.8, 2.0)	McGuyer et al. [12]
$1_u(-1, 1)$	353.236(35)	0	X	−0.01570(5)*	(−50, 50)	McGuyer et al. [11]
		1	0.8751(121)	−0.0108(14)	(−4.0, 1.0)	McGuyer et al. [10]
$0_u(-4, 1)$	1084.093(33)	0	X	−0.0026(10)	(−4.0, 1.0)	McGuyer et al. [10]
		1	0.1994(28)	−0.0058(4)	(−4.0, 1.0)	McGuyer et al. [10]
$1_g(-3, 1)$	1669(1)	0	X	−0.01143(8)	(−2.0, −2.0)	McGuyer et al. [12]
		1	1.046(1)	−0.0087(5)	(−1.8, 2.0)	McGuyer et al. [12]
$1_u(-2, 1)$	2683.722(32)	0	X	−0.0112(25)	(−4.0, 1.0)	McGuyer et al. [10]
		1	0.8171(113)	−0.0047(15)	(−4.0, 1.0)	McGuyer et al. [10]
$0_u(-5, 1)$	3463.280(33)	0	X	−0.0022(11)	(−4.0, 1.0)	McGuyer et al. [10]
		1	0.2703(40)	−0.0040(7)	(−4.0, 1.0)	McGuyer et al. [10]
$1_g(-4, 1)$	5168(2)	0	X	−0.00765(6)	(−1.8, 2.0)	McGuyer et al. [12]
		1	1.045(1)	−0.0066(7)	(−1.8, 2.0)	McGuyer et al. [12]
$1_u(-3, 1)$	8200.163(39)	0	X	−0.0157(17)	(−4.0, 1.0)	McGuyer et al. [10]
		1	0.2085(31)	−0.0056(5)	(−4.0, 1.0)	McGuyer et al. [10]
$0_u(-6, 1)$	8429.650(42)	0	X	−0.0011(8)	(−4.0, 1.0)	McGuyer et al. [10]
		1	1.3037(178)	−0.0084(18)	(−4.0, 1.0)	McGuyer et al. [10]

See text for details

Table 4.2 Linear and quadratic Zeeman shifts for all singly-excited rovibrational levels of ^{88}Sr$_2$ with $E < 8.0$ GHz and $J' = 2$

$J' = 2$						
	E (MHz)	$|m|$	β_1 (MHz/G)	β_2 (MHz/G^2)	(B_i, B_f) (G)	Ref.
$1_g(-1, 2)$	7(1)	0	X	0.0423(14)	(−1.8, 1.8)	McGuyer et al. [12]
		1	0.5990(15)	0.0189(14)	(−1.8, 1.8)	McGuyer et al. [12]
		2	1.1854(29)	−0.0544(25)	(−1.8, 1.8)	McGuyer et al. [12]
$1_g(-2, 2)$	270(1)	0	X	0.01418(73)	(−1.8, 2.0)	McGuyer et al. [12]
		1	0.4882(13)	0.0058(15)	(−1.8, 2.0)	McGuyer et al. [12]
		2	0.9720(14)	0.0123(17)	(−1.8, 2.0)	McGuyer et al. [12]
$1_u(-1, 2)$	287(1)	0	X	0.01220(8)*	(−50, 50)	McGuyer et al. [11]
		1	0.3471(5)*	0.00839(2)*	(−38, 38)	McGuyer et al. [11]
		2	0.6939(8)	−0.0025(1)	(−14, 14)	McGuyer et al. [11]
$1_g(-3, 2)$	1581(1)	0	X			
		1	0.434(4)	0.002(7)	(1.8, 2.0)	McGuyer et al. [12]
		2	0.855(3)	−0.002(3)	(1.8, 2.0)	McGuyer et al. [12]
$1_u(-2, 2)$	2569(1)	0	X			
		1	0.363(2)	0.006(1)	(−4.0, 1.0)	McDonald [9]
		2	0.725(4)	−0.004(2)	(−4.0, 1.0)	McDonald [9]
$1_g(-4, 2)$	5035(1)	0	X			
		1	0.4128(13)	0.00124(12)	(−18, 18)	McGuyer et al. [12]
		2	0.823(2)	−0.005(2)	(−1.8, 2.0)	McGuyer et al. [12]

See text for details. Note that all measurements assume negligible Zeeman shift of the ground state, an assumption which could introduce errors in the recorded linear Zeeman shifts for $|m| \geq 2$ at the few hundred Hz/G level

4.2 Linear Zeeman Shifts (Low-Field)

Linear Zeeman shifts are due to interactions between the angular momentum of charged particles and an externally applied magnetic field. For ^{88}Sr, which lacks nuclear spin, the Zeeman shift is due entirely to interactions with the electronic angular momentum, defined by the Hamiltonian

$$\hat{H}_Z = \hat{\vec{\mu}} \cdot \vec{B} = \mu_B(g_S\hat{S} + g_L\hat{L}) \cdot \vec{B}, \tag{4.2}$$

where $g_L = 1$ and $g_S \approx 2$ are the electron orbital and spin g-factors, respectively. The linear Zeeman shift is then defined simply as the first-order perturbation induced by this perturbing Hamiltonian:

$$\Delta E = \langle \Psi^0 | \hat{H}_Z | \Psi^0 \rangle \tag{4.3}$$

4.2 Linear Zeeman Shifts (Low-Field)

Table 4.3 Linear and quadratic Zeeman shifts for all singly-excited rovibrational levels of $^{88}\text{Sr}_2$ with $E < 2.3\,\text{GHz}$ and $J' = 3$

$J' = 3$						
	E (MHz)	$|m|$	β_1 (MHz/G)	β_2 (MHz/G^2)	(B_i, B_f) (G)	Ref.
$0_u(-2, 3)$	0.626(12)	0	X	−0.1820(67)	(−1.0, 1.0)	McGuyer et al. [10]
		1	0.3835(55)	−0.1421(68)	(−1.0, 1.0)	McGuyer et al. [10]
		2	0.7525(101)	−0.1326(80)	(−1.0, 1.0)	McGuyer et al. [10]
		3	1.1256(150)	−0.1256(62)	(−1.0, 1.0)	McGuyer et al. [10]
$0_u(-3, 3)$	132	0	X	−0.0058(21)	(−3.8, 1.0)	McGuyer et al. [10]
		1	0.2457(36)	−0.0060(7)	(−3.8, 1.0)	McGuyer et al. [10]
		2	0.4855(69)	−0.0086(13)	(−3.8, 1.0)	McGuyer et al. [10]
		3	0.7098(96)	−0.0084(20)	(−3.8, 1.0)	McGuyer et al. [10]
$1_u(-1, 3)$	171(1)	0	X	−0.00638(6)	(−17, 20)	McDonald [9]
		1	0.04715(61)	−0.00599(4)*	(−17, 20)	McDonald [9]
		2	0.09438(45)	−0.00480(2)*	(−17, 20)	McDonald [9]
		3	0.14228(60)	−0.00255(3)*	(−17, 20)	McDonald [9]
$1_g(-2, 3)$	193(1)	0	X			
		1	0.1473(26)		(−2.3, 1.0)	McDonald [9]
		2	0.2946(17)		(−2.3, 1.0)	McDonald [9]
		3	0.4485(37)		(−2.3, 1.0)	McDonald [9]
$0_u(-4, 3)$	901.0(5)	0	X	−0.00242(35)	(−1.0, 4.0)	McDonald [9]
		1	0.1806(24)	−0.0014(9)	(−1.0, 4.0)	McDonald [9]
		2	0.36369(87)	−0.00293(33)	(−1.0, 4.0)	McDonald [9]
		3				
$1_g(-3, 3)$	1438(1)	0	X			
		1	0.1880(56)		(0.0, 6.1)	McDonald [9]
		2	0.3550(38)		(0.0, 6.1)	McDonald [9]
		3	0.5291(28)		(0.0, 6.1)	McDonald [9]

See text for details. Note that all measurements assume negligible Zeeman shift of the ground state, an assumption which could introduce errors in the recorded linear Zeeman shifts for $|m| \geq 2$ at the few hundred Hz/G level

4.2.1 Calculation of the Linear Zeeman Shift of the $^3P_1$ State of ^{88}Sr

In labelling the $^3P_1$ state as we have, we are implicitly assuming *Russell-Saunders coupling*, i.e. that spin-orbit coupling is small and that the applied magnetic fields are weak. In this case, the atomic wavefunctions are assumed to be eigenstates of $\vec{L}^2$, $\vec{S}^2$, and $\vec{J}^2$, where the total electronic angular momentum $\vec{J} = \vec{L} + \vec{S}$ has a projection along the quantization axis of m_j. The linear Zeeman shift for such a system is given by the well-known formula [20]:

Table 4.4 Linear and quadratic Zeeman shifts for all singly-excited rovibrational levels of ^{88}Sr$_2$ with $E < 1.2$ GHz and $J' = 4$

$J' = 4$						
	E (MHz)	$\|m\|$	β_1 (MHz/G)	β_2 (MHz/G^2)	(B_i, B_f) (G)	Ref.
$1_u(-1, 4)$	56.2(1.0)	0	X	0.01234(2)*	$(-40, 40)$	McGuyer et al. [11]
		1	0.1041(4)*	0.01163(2)*	$(-40, 40)$	McGuyer et al. [11]
		2	0.2030(9)	0.0087(1)	$(-13, 13)$	McGuyer et al. [11]
		3				
		4				
$1_g(-2, 4)$	114(1)	0	X	$-0.009(14)$	$(-0.15, 0.35)$	McDonald [9]
		1	0.2477(30)	$-0.038(29)$	$(-0.15, 0.35)$	McDonald [9]
		2	0.4934(51)	0.024(24)	$(-0.15, 0.35)$	McDonald [9]
		3				
		4				

See text for details. Note that all measurements assume negligible Zeeman shift of the ground state, an assumption which could introduce errors in the recorded linear Zeeman shifts for $|m| \geq 2$ at the few hundred Hz/G level

$$\Delta E = \mu_B |B| m_j \left[g_L \frac{j(j+1) + l(l+1) - s(s+1)}{2j(j+1)} + g_S \frac{j(j+1) - l(l+1) + s(s+1)}{2j(j+1)} \right]. \tag{4.4}$$

The $^3P_1$ state carries $s = 1$, $l = 1$, and $j = 1$. Assuming that $g_e = 2$, we get the following result for the total linear Zeeman shift:

$$\frac{\Delta E}{|B|} = \frac{3}{2} \mu_B m_j \approx 2.0994 \ldots \cdot (\mathbf{m_j}) \text{ MHz/G} \tag{4.5}$$

The Zeeman shift calculated above has been used to calibrate all of our coils, and therefore all Zeeman shifts reported in this thesis are defined with respect to this calibration. However, note that even if Russel-Saunders coupling is a perfectly accurate description for the $^3P_1$ state, we would expect our calculated Zeeman shift to be wrong at approximately the part-per-thousand level due to the fact that g_e is not exactly 2, but rather equal to 2.002 319 304 361 82(52) [6, 14].

4.2.2 'Ideal' Linear Zeeman Shifts for Molecules Satisfying Hund's Case (c)

Calculating the linear Zeeman shift in our Hund's case (c) basis is complicated by the fact that while the magnetic field defined in Eq. (4.2) is defined with respect to an external "laboratory frame," the labels of our basis functions $|\eta, J_a; J, \Omega, M_J\rangle$ are defined with respect to the (rotating) molecular frame. To transform between these

4.2 Linear Zeeman Shifts (Low-Field)

Table 4.5 Zeeman shift coefficients for up to B^6 field dependence are given for $1_u(v=-1, J=1-4)$

	$|m|$	β_1 (MHz/G)	β_2 (MHz/G^2)	β_3 (MHz/G^2)	β_4 (MHz/G^2)	β_5 (MHz/G^2)	β_6 (MHz/G^2)
$1_u(-1,1)$	0	X	−0.01570(5)	X	2.77(7)E−6	X	−3.84(2)E−10
	1	0.8751(121)	−0.0108(14)	N/A	N/A	N/A	N/A
$1_u(-1,2)$	0	X	0.0122(8)	X	−3.33(8)E−6	X	3.8(2)E−10
	1	0.3471(5)	0.00839(2)	1.51(14)E−5	−1.36(2)E−6	7.9(8)E−9	N/A
	2	0.6939(8)	−0.0025(1)	N/A	N/A	N/A	N/A
$1_u(-1,3)$	0	X	−0.00638(6)	X	N/A	X	N/A
	1	0.04715(61)	−0.00599(4)	4.7(3)E−5	N/A	N/A	N/A
	2	0.09438(45)	−0.00480(2)	6.98(22)E−5	N/A	N/A	N/A
	3	0.14228(60)	−0.00255(3)	3.94(30)E−5	N/A	N/A	N/A
$1_u(-1,4)$	0	X	0.01234(2)	X	−1.62(2)E−6	X	N/A
	1	0.1041(4)	0.01163(2)	3.42(13)E−5	−1.53(2)E−6	1.03(9)E−8	N/A
	2	0.2030(9)	0.0087(1)	N/A	N/A	N/A	N/A
	3	N/A	N/A	N/A	N/A	N/A	N/A
	4	N/A	N/A	N/A	N/A	N/A	N/A

Entries marked "X" must be zero by symmetry, and entries marked "N/A" were forced to equal zero either because measurements were not precise enough to resolve a clear need for these coefficients to be incorporated in order to summarize the data, or because the sublevel in question was not observed. Note that all measurements assume negligible Zeeman shift of the ground state, an assumption which could introduce errors in the recorded linear Zeeman shifts for $|m| \geq 2$ at the few hundred Hz/G level. This data comes from the same measurements used to create Tables 4.1, 4.2, 4.3, and 4.4

frames we can make use of *spherical tensor algebra* (see, e.g., Chapter 5 of Brown and Carrington [2], and a closely related calculation of linear Zeeman shifts of the long-range He...Ar$^+$ molecular ion in Ref. [3]).

In spherical tensor notation, Eq. (4.2) has the form:

$$\hat{H}_Z = g_S \mu_B T^1(\vec{B}) \cdot T^1(\hat{S}) + g_L \mu_B T^1(\vec{B}) \cdot T^1(\hat{L}). \quad (4.6)$$

Assuming the magnetic field is directed along the $p = 0$ (i.e., "z") axis, while $\hat{S}$ and $\hat{L}$ are defined with respect to the internuclear axis, their dot products can be rewritten with *Wigner rotation matrices*:

$$\hat{H}_Z = g_S \mu_B T_0^1(\vec{B}) \sum_q \mathscr{D}_{0q}^{(1)}(\omega)^* T_q^1(\hat{S}) + g_L \mu_B T_0^1(\vec{B}) \sum_q \mathscr{D}_{0q}^{(1)}(\omega)^* T_q^1(\hat{L}) \quad (4.7)$$

The process for solving for the energy perturbation due to each of the above terms is practically identical, and so for the rest of this calculation I'll focus only on the first. The result for the second follows directly by analogy.

First, note that the total electronic angular momentum J_a is decoupled from the total angular momentum J, i.e. that $\hat{J}_a \cdot \hat{J}$ is small compared to $\hat{J}_a^2$ and $\hat{J}^2$. Then, because $\hat{S}$ acts on the electronic angular momentum and $\mathscr{D}_{0q}^{(1)}(\omega)^*$ is a function of the basis in which the molecular coordinates are written, we can rewrite the first order perturbation of the first part ΔE_1 as:

$$\Delta E_1 = \left\langle \eta(\Omega), J_a; J, \Omega, M_J \middle| g_S \mu_B T_0^1(\vec{B}) \sum_q \mathscr{D}_{0q}^{(1)}(\omega)^* T_q^1(\hat{S}) \middle| \eta'(\Omega'), J_a'; J', \Omega', M_{J'} \right\rangle \quad (4.8)$$

$$= g_S \mu_B B_z \sum_q \underbrace{\langle J, \Omega, M_J | \mathscr{D}_{0q}^{(1)}(\omega)^* | J', \Omega', M_{J'} \rangle}_{A} \underbrace{\langle J_a, \Omega | T_q^1(\hat{S}) | J_a', \Omega' \rangle}_{B} \quad (4.9)$$

where I have chosen $(\eta(\Omega), J_a, J, \Omega, M_J) \neq (\eta'(\Omega'), J_a', J', \Omega', M_{J'})$ for now for completeness. For the ideal Hund's case (c) result, we will eventually set initial and final quantum numbers equal to one another.

Written in this way, we can see that there are two components of this equation which can be solved individually.

Part A From Brown and Carrington [2] Equation 5.184, we find:

$$\langle J, \Omega, M_J | \mathscr{D}_{0q}^{(1)}(\omega)^* | J', \Omega', M_{J'} \rangle = (-1)^{M_J - \Omega} \sqrt{[J][J']} \begin{pmatrix} J & 1 & J' \\ -M_J & 0 & M_{J'} \end{pmatrix} \begin{pmatrix} J & 1 & J' \\ -\Omega & q & \Omega' \end{pmatrix}, \quad (4.10)$$

where I have adopted the shorthand $[J] = 2J + 1$ for the sake of clarity, and the terms in parentheses represent "Wigner 3j symbols."

4.2 Linear Zeeman Shifts (Low-Field)

We are most interested in the first-order perturbation to an ideal Hund's case (c) state, which in the above expression requires that the initial and final quantum numbers are equal. Doing so, and using the following result from Appendix C of Brown and Carrington,

$$\begin{pmatrix} J & 1 & J \\ -m & q & m \end{pmatrix} = \delta_{q0}(-1)^{3J+m} \frac{m}{\sqrt{J(J+1)[J]}}, \quad (4.11)$$

yields the following result:

$$\langle J, \Omega, M_J | \mathcal{D}_{0q}^{(1)}(\omega)^* | J, \Omega, M_J \rangle = (-1)^{M_J - \Omega}[J] \begin{pmatrix} J & 1 & J \\ -M_J & 0 & M_J \end{pmatrix} \begin{pmatrix} J & 1 & J \\ -\Omega & q & \Omega \end{pmatrix} \quad (4.12)$$

$$= (-1)^{2M_J + 6J} \frac{M_J \Omega}{J(J+1)} \delta_{q0} \quad (4.13)$$

Part B To evaluate the next component of Eq. (4.9), we again turn to Brown and Carrington. Using Equation 5.172 (i.e., the Wigner-Eckart theorem) we find:

$$\langle J_a, \Omega | T_q^1(\hat{S}) | J_a', \Omega' \rangle = (-1)^{J_a - \Omega} \begin{pmatrix} J_a & 1 & J_a' \\ -\Omega & q & \Omega' \end{pmatrix} \langle J_a || T_q^1(\hat{S}) || J_a' \rangle, \quad (4.14)$$

where $\langle J_a || T_q^1(\hat{S}) || J_a' \rangle$ is a *reduced matrix element*.

To evaluate this reduced matrix element, we need to be careful. The operator $\hat{S}$ acts on only part of the operator $\hat{J}_a = \hat{S} + \hat{L}$. Rewriting $|J_a\rangle$ as $|L, S, J_a\rangle$ and using Equation 5.175 of Brown and Carrington yields:

$$\langle J_a || T_q^1(\hat{S}) || J_a' \rangle = \langle L, S, J_a || T_q^1(\hat{S}) || L', S', J_a' \rangle \quad (4.15)$$

$$= \delta_{LL'}(-1)^{J_a + L + 1 + S'} \sqrt{[J_a][J_a']} \begin{Bmatrix} S' & J_a' & L \\ J_a & S & 1 \end{Bmatrix} \langle S || T^1(\hat{S}) || S' \rangle, \quad (4.16)$$

where the term in brackets is a *Wigner 6j symbol*.

The final inner product is a *truly* reduced matrix element, and can be evaluated with Brown and Carrington Equation 5.179:

$$\langle S || T^1(\hat{S}) || S' \rangle = \delta_{SS'} \sqrt{S(S+1)[S]} \quad (4.17)$$

Putting everything together results in a monster expression which can be reduced once again by setting all initial and final quantum numbers equal to one another:

$$\langle J_a, \Omega | T_q^1(\hat{S}) | J_a, \Omega \rangle = (-1)^{J_a-\Omega} \begin{pmatrix} J_a & 1 & J_a \\ -\Omega & q & \Omega \end{pmatrix}$$

$$\times (-1)^{J_a+L+1+S}[J_a] \begin{Bmatrix} S & J_a & L \\ J_a & S & 1 \end{Bmatrix} \times \sqrt{S(S+1)[S]} \quad (4.18)$$

We need one more identity from Brown and Carrington (Appendix D), which states:

$$\begin{Bmatrix} S & J_a & L \\ J_a & S & 1 \end{Bmatrix} = (-1)^{J_a+S+L+1} \frac{S(S+1) + J_a(J_a+1) - L(L+1)}{\sqrt{S(2S+1)(2S+2)(J_a)(2J_a+1)(2J_a+2)}}$$

(4.19)

Carefully adding all contributions yields the final result for part B:

$$\langle J_a, \Omega | T_q^1(\hat{S}) | J_a', \Omega' \rangle = (-1)^{6J_a+2L+2S+2} \delta_{q0} \Omega \frac{S(S+1) + J_a(J_a+1) - L(L+1)}{2J_a(J_a+1)}$$

(4.20)

Total Linear Zeeman Shift Assuming all angular momentum quantum numbers take integer values (as is true in our case), the complicated factors of $(-1)^n$ disappear. We then recognize that if we were to repeat the calculation for the *orbital* angular momentum $\hat{L}$ component, then the effect would be to replace g_S with g_L, as well as to swap the positions of L and S in the final expression. Combining everything then yields a final result for the total linear Zeeman shift in a Hund's case (c) molecule:

$$\boxed{\langle \hat{H}_Z \rangle = \mu_B B_z \frac{\Omega^2 M_J}{2J(J+1)} \left((g_S + g_L) + (g_S - g_L) \frac{S(S+1) - L(L+1)}{J_a(J_a+1)} \right)}$$

(4.21)

(Note that a slightly less general version of this formula was also given in [19].)

We're strictly interested in only two "flavors" of molecule for the purposes of this thesis: those consisting of two atoms in each in the $^1S_0$ state, and those for which one of the atoms is in the $^3P_1$ state. The linear Zeeman shift for electronic ground-state molecules is clearly zero, since all electronic angular momentum quantum numbers are equal to zero. In the excited $^3P_1$ state, we have $S = 1$, $L = 1$, and $J_a = 1$. Plugging these values into Eq. (4.21), as well as $g_L = 1$ and $g_S = 2$, yields the following predictions for the Zeeman shifts for molecules with various total angular momenta J and $\Omega = 1$ ($\Omega = 0$ yields zero linear Zeeman shift for all J):

- $J = 1$: $\Delta E = 1.0497 \times M_J$ (MHz/G)
- $J = 2$: $\Delta E = 0.3499 \times M_J$ (MHz/G)
- $J = 3$: $\Delta E = 0.17495 \times M_J$ (MHz/G)
- $J = 4$: $\Delta E = 0.10497 \times M_J$ (MHz/G)

4.2 Linear Zeeman Shifts (Low-Field)

How well do these predictions match our data? Tables 4.1, 4.2, 4.3, and 4.4 show all measurements of linear Zeeman shift coefficients made to date by our lab. While there is clearly a large spread in the measured Zeeman shifts for different levels, a pattern can be identified (highlighted in green in the table) which seems to do a good job of predicting when the measurement will match the calculation of the ideal Hund's case (c) prediction: **If J is *odd* and the wavefunction symmetry is *gerade*, or if J is *even* and the wavefunction symmetry is *ungerade*, then linear Zeeman shifts will be close to "ideal."**

What determines whether or not a particular rovibrational level will adhere to the ideal Hund's case (c) prediction? The answer lies in investigating our original assumption about whether Ω was a valid quantum number.

4.2.2.1 Aside: Note on the Impact of Using Pure Ω States vs Parity-Adapted Eigenstates

As we discovered in Chap. 2, states with the 1_u label are actually linear superpositions of eigenfunctions with $\Omega = \pm 1$. However, in deriving ideal predictions for linear Zeeman shifts, we assumed Ω to take a single value. Will this impact our predictions for what should be observed in experiment? For linear Zeeman shifts it turns out that the answer is *no*, since the total shift is proportional to Ω^2 and independent of the sign of Ω. However, we *will* need to take into account the full form of the "1_u" basis states in order to accurately describe Coriolis coupling.

4.2.2.2 Coriolis Coupling and Mixing Angles

Coriolis coupling, i.e. a coupling between the vibration and rotation of a molecule, can cause the unperturbed rovibrational levels in a molecule to become linear superpositions of states with $\Delta\Omega = \pm 1$. There's no simple rule for predicting the degree of mixing for a particular rovibrational level when such mixing is allowed, but this quantity can be calculated numerically with sufficient knowledge of the shapes of molecular potentials and their interactions [1, 10]. And since Coriolis coupling can cause deviations of linear Zeeman shifts from the ideal Hund's case (c) predictions, we can invert the problem and use precise measurements of linear Zeeman shifts to characterize the degree to which particular rovibrational levels are mixed.

Instead of assuming that an observed rovibrational level is a pure $|\Omega| = 0$ or 1 state, let's instead let an observed state $|v, J, M_J\rangle$ be a superposition of ideal Hund's case (c) states, e.g.

$$|v, J, M_J\rangle = \cos(\theta)|v(0), J, M_J\rangle + \sin(\theta)|v(1), J, M_J\rangle, \quad (4.22)$$

where the component wavefunctions $|v(\Omega), J, M_J\rangle$ are defined by

$$|v(0), J, M_J\rangle = |\eta_{|\Omega|=0}; v(|\Omega|=0)\rangle |J, \Omega = 0, M_J\rangle \qquad (4.23)$$

$$|v(1), J, M_J\rangle = |\eta_{|\Omega|=1}; v(|\Omega|=1)\rangle \left(\frac{1}{\sqrt{2}}|J, \Omega = +1, M_J\rangle + \frac{1}{\sqrt{2}}|J, \Omega = -1, M_J\rangle\right), \qquad (4.24)$$

where $|J, \Omega, M_J\rangle$ is a purely rotational state, and $|\eta_{|\Omega|}; v(|\Omega|)\rangle$ represents the electronic and vibrational parts of the wavefunction.

According to the above definition, states for which $\theta = 0°$ represent pure 0_u states, while states for which $\theta = 90°$ represent pure 1_u states. Using Eq. (4.22) to define our wavefunction, we can then calculate the linear Zeeman shift, and relate this quantity to the purity of our state through the mixing angle θ. We therefore find:

$$\begin{aligned}\Delta E &= \langle v, J, M_J | \hat{H}_Z | v, J, M_J\rangle \\ &= \cos^2\theta \langle J, \Omega = 0, M_J | \hat{H}_Z | J, \Omega = 0, M_J\rangle \\ &+ \frac{\sin^2\theta}{2}\Big(\langle J, \Omega = 1, M_J | \hat{H}_Z | J, \Omega = 1, M_J\rangle + \langle J, \Omega = -1, M_J | \hat{H}_Z | J, \Omega = -1, M_J\rangle\Big) \\ &+ \frac{\sin 2\theta \langle v(0)|v(1)\rangle}{\sqrt{2}}\Big(\langle \eta_0; J, \Omega = 0, M_J | \hat{H}_Z | \eta_1; J, \Omega = 1, M_J\rangle \\ &+ \langle \eta_0; J, \Omega = 0, M_J | \hat{H}_Z | \eta_1; J, \Omega = -1, M_J\rangle\Big).\end{aligned} \qquad (4.25)$$

Note that only terms satisfying $\Delta\Omega = 0, \pm 1$ are present (i.e., there is no mixing between $\Omega = -1, +1$), which is required because of a selection rule for the Zeeman Hamiltonian. We've already calculated the first two terms, since they represent simply the first-order Zeeman shifts for ideal Hund's case (c) states. The third term can be calculated in the same way, with the only difference being that Ω, Ω' are not forced to be equal.

Solving for this third term (which for clarity is not repeated here) and adding everything together yields the linear Zeeman shift as a function mixing angle θ between $\Omega = 0, 1$ states:

$$\Delta E = \frac{3}{2} M_J \mu_B B \left(\frac{\sin^2\theta}{J(J+1)} + \frac{\sin 2\theta}{\sqrt{J(J+1)}}\langle v(0)|v(1)\rangle\right). \qquad (4.26)$$

In our calculation of mixing angles from experimental Zeeman shift data, we assumed the vibrational wavefunction overlap $\langle v(0)|v(1)\rangle$ was exactly equal to 1, an approximation which was supported by numerical calculations from our collaborators. Later work [1] improved agreement between experiment and theory partly by making more accurate calculations of this overlap.

4.2.3 Validity of Hund's Case (c)

We're now in a position to understand why only odd-J gerade states and even-J ungerade states adhere nearly perfectly to the ideal Hund's case (c) prediction. When both $\Omega = 0$ and $|\Omega| = 1$ states are possible for a given combination of parity and rotational angular momentum, Coriolis coupling will cause strong mixing between the 0_u and 1_u potentials, and the observed levels will have Zeeman shifts determined by a competition between the ideal 0_u shift (i.e. zero) and the ideal 1_u shift given by Eq. (4.21). If, however, $\Omega = 0$ is forbidden by quantum statistics, then the observed levels must be pure 1_u states, and therefore adhere nearly perfectly to the ideal Hund's case (c) prediction.

It is *pretty neat* that within one molecule, and between rovibrational levels differing only by one unit of rotational angular momentum, we can see both the validation and breakdown of the Hund's case (c) model.

4.3 Quadratic (and Higher Order) Zeeman Shifts

Quadratic Zeeman shifts are the result of second-order perturbations of molecular binding energies due to the presence of a magnetic field. The second order correction $E_k^{(2)}$ to the unperturbed energy $E_k^{(0)}$ of a state $|k\rangle$ is given by the well-known formula

$$\Delta E_k^{(2)} = \sum_{k \neq n} \frac{|\langle k^{(0)}|\hat{H}_Z|n^{(0)}\rangle|^2}{E_n^{(0)} - E_k^{(0)}}, \qquad (4.27)$$

where the perturbing Hamiltonian $\hat{H}_Z$ is in this case the Zeeman Hamiltonian, which connects states with $\Delta J = 0, 1$ and $\Delta M_J = 0$ (but $\Delta J \neq 0$ if $J = 0$).

It is clear that the sum in Eq. (4.27) means that any exact calculation of the expected shift will be extremely complicated. However, we can develop an intuitive understanding of how these shifts relate to the structure of the molecule by making a few simplifying assumptions. There are (at least) two different ways we can proceed, which give different pictures about the underlying physics responsible for these shifts.

4.3.1 Option 1: Coriolis Coupling of the $\Omega = 0, 1$ Potentials

As was discussed in Sect. 4.2.2.2, the states we observe in singly-excited [88]Sr_2 are not perfectly ideal Hund's case (c) eigenfunctions. Instead, they are superpositions of $\Omega = 0$ and $\Omega = 1$ eigenstates given by Eqs. (4.23) and (4.24). This realization implies that the dominant contribution to the perturbation sum might be due to the Zeeman Hamiltonian coupling states of different Ω.

From our knowledge of the form of the Zeeman Hamiltonian $\hat{H}_Z$ and the Coriolis mixed wavefunctions described by (4.23) and (4.24), it's clear that the numerator $|\langle k^{(0)}|\hat{H}_Z|n^{(0)}\rangle|^2$ of Eq. (4.27) will take the form:

$$\text{(numerator)} = \mu_B^2 \cdot B^2 \cdot |\langle v(0)|v(1)\rangle|^2 \times (\text{function of } J, M_J, \Omega \ldots). \quad (4.28)$$

Numerical calculations show that $\langle v(0)|v(1)\rangle$ is of order unity (though recent work has modeled this wavefunction overlap more accurately [1]). The "function of $J, M_J, \Omega \ldots$" takes a bit more work (i.e., spherical tensor algebra), but can also be shown to be approximately of order unity. Therefore the behavior of this perturbation will be mainly determined by the denominator.

The denominator is the difference in energies of the two states being coupled by the Zeeman Hamiltonian. Since we're assuming that Coriolis coupling is the dominant contributor here, let's approximate this energy difference simply as the difference in energies between a state of radius R confined to the 0_u (or 0_g) potential vs one confined to the 1_u (or 1_g) potential. The shapes of the 1_u and 0_u potentials are well-known [1, 16, 18, 22]. At long range, the electronic parts of the 1_u and 0_u potentials can be modeled primarily by van der Waals ($V \propto \frac{C_6}{R^6}$) and dipole–dipole ($V \propto \frac{C_3}{R^3}$) interactions, where C_6 and C_3 are coefficients determined primarily by atomic properties. They can be approximated with the following Equations [22]:

$$V_{0_u}^{\text{electronic}} = -C_{6,0_u}/R^6 - 2C_3/R^3 \quad (4.29)$$

$$V_{1_u}^{\text{electronic}} = -C_{6,1_u}/R^6 + C_3/R^3 \quad (4.30)$$

The coefficients $C_{6,0_u}$ and $C_{6,1_u}$ are very nearly equal (to within $\sim 6\%$ according to recent calculations [16]), but there is a sign and factor of 2 difference in the $1/R^3$ term which dominates the difference between these two potentials. (This positive $1/R^3$ term accounts for the repulsive bump in the 1_u potential seen, e.g., in Fig. 5.2.) If in Eq. (4.27) we then approximate the denominator $E_n^{(0)} - E_k^{(0)} \approx V_{1_u} - V_{0_u}$, and assuming that the largest contribution to the perturbation comes from this Coriolis mixing, we find:

$$\Delta E_k^{(2)} = \cdot \frac{\mu_B^2 \cdot B^2 \cdot (\text{function of } J, M_J, \Omega \ldots)}{(-C_{6,1_u}/R^6 + C_{6,0_u}/R^6) + (C_3/R^3 + 2C_3/R^3)}$$

$$\approx \mu_B^2 B^2 \frac{f(J, M_J, \Omega) R^3}{3 C_3}, \quad (4.31)$$

where $f(J, M_J, \Omega)$ is a function which can be determined via angular momentum algebra.

According to this argument, **the quadratic Zeeman shift coefficients β_2 described in Tables 4.1, 4.2, 4.3, and 4.4 should scale with bond length cubed**. In other words, as a molecule gets bigger, the magnetic field-induced second-order perturbations to its energy levels should increase as well.

4.3 Quadratic (and Higher Order) Zeeman Shifts

Figure 4.1 shows the quadratic Zeeman shift coefficients β_2 plotted vs bond length R for all $J' = 1$ levels observed so far, with $M_J = 0$ in panel (a) and $|M_J| = 1$ in panel (b). The dashed lines represent fits to the data with the function $\beta_2 \propto R^3$, while the solid lines represent another fit to be discussed in the next section. And while the dashed line fit does do a fair job of *qualitatively* describing the data, it clearly disagrees with the most accurate measurements taken (i.e., data for the 1_g, $M_J = 0$ states). Since this is only a hand-wavy approximation, we might be satisfied with that level of agreement; this approach is described in [10]. But it turns out that we can make another valid approximate argument which reproduces the data even better.

4.3.2 Option 2: Decreased Level Spacing Near the Top of the Potential

The shape of an energy potential determines the locations of bound states within it. The canonical example of the harmonic oscillator potential possesses energy levels whose spacing is constant no matter how high above the ground state one goes (see, e.g., Fig. 7.2a). For molecular potentials, however, which rise steeply at intermediate bond lengths but taper off as $R \to \infty$, the levels become more tightly bunched as one approaches dissociation. Clearly the level spacing will be important for determining the second order energy perturbations, since the denominator of Eq. (4.27) is precisely this spacing. If we can relate the level spacing in a simple way to the shape of the potential, we'll have another conceptual tool for thinking about how quadratic Zeeman shifts are related to structure.

4.3.2.1 Applying the LeRoy-Berstein Formula

If a potential can be approximately described at long range by the formula $V(R) = D - C_n/R^n$, where D is the dissociation energy, then the binding energy of the vth level $E(v)$ will be approximately given by what's known as the "LeRoy-Bernstein formula":

$$E(v) \approx -[(v_D - v)H_n]^{\frac{2n}{n-2}}, \qquad (4.32)$$

where H_n is a function of n and various constants, v_D is the "effective" vibrational number for a state bound at the dissociation limit, and $E(v)$ represents the rovibrational level energy minus the dissociation threshold [8].

This formula is helpful because it connects the vibrational number v to the molecular bond length R. To see why this is useful, let's first assume that the sum in Eq. (4.27) is dominated by mixing with nearby states. The denominator of the largest term is then just the difference between adjacent levels, or:

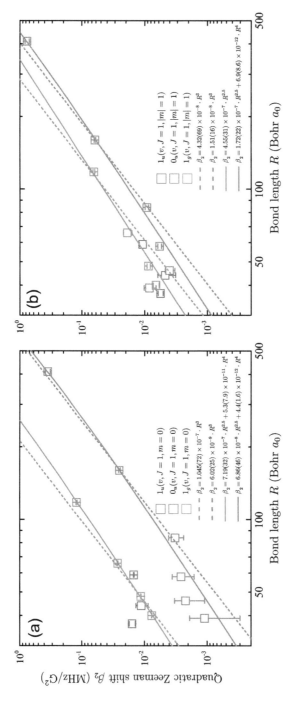

Fig. 4.1 Magnitudes of the quadratic Zeeman shift coefficients β_2 from Table 4.1 are plotted vs bond length for (**a**) $M_J = 0$ and (**b**) $|M_J| = 1$. Dashed lines indicate fits assuming perturbation dominated by the difference in shapes between the $1_{u/g}$ and $0_{u/g}$ potentials, whereas solid lines assume the R-dependence of β_2 comes from the behavior of the level spacing as one walks up the potential

4.3 Quadratic (and Higher Order) Zeeman Shifts

$$E_k - E_\mu = \Delta E \approx \frac{\partial E}{\partial v} \Delta v = \frac{\partial E}{\partial v}, \tag{4.33}$$

where the final equality is because $\Delta v = 1$ for the nearest neighbor. The LeRoy-Bernstein formula then allows us to write $\frac{\partial E}{\partial v}$ in terms of R by differentiating Eq. (4.32) with respect to v:

$$\frac{\partial E}{\partial v} \propto E_v^{\frac{n+2}{2n}} \propto \frac{1}{R^{\frac{n+2}{2}}} \tag{4.34}$$

If, as in Sect. 4.3.1, we then assume that the numerator of Eq. (4.27) is of order unity (or at least roughly independent of vibrational number and bond length), then we substitute $E_k - E_\mu \approx \frac{\partial E}{\partial v}$ to obtain the result:

$$\Delta E_k^{(2)} \propto \mu_B^2 \cdot B^2 \cdot \left(\frac{\partial E}{\partial v}\right)^{-1} \propto \mu_B^2 \cdot B^2 \cdot R^{\frac{n+2}{2}} \tag{4.35}$$

At very large bond lengths, the largest term in the potential $V(R)$ will be of the form $\propto \frac{C_3}{R^3}$, **and so for very large molecules we would expect $\beta_2 \propto R^{2.5}$**. At smaller bond lengths, when $V(R)$ is dominated by the $\propto \frac{C_6}{R^6}$ term, we would expect $\beta_2 \propto R^4$.

In Fig. 4.1, solid lines represent fits to the data of the form $\beta_2 = A \cdot R^{2.5} + B \cdot R^4$. Though A and B were both left as free parameters, in all four fits the coefficient B is within a few standard deviations of zero. For nearly every series shown this fit is better than the dashed line fit, because the slope of the R^3 on the log-log plot is clearly a little too steep. The agreement for the $1_g(J = 1, M_J = 0)$ data is particularly striking.

Note that the above result is directly related to another result which will be discussed later, i.e. that the predissociative linewidths of weakly-bound subradiant states are proportional to the level spacing. In the case of linewidths, R^4 dependence seemed to prevail over nearly the entire experimental range, whereas here the data is well-summarized by $R^{2.5}$ dependence. Why this difference might be isn't immediately clear, but would certainly be worthy of future follow-up.

4.3.3 Comparison of Atomic and Molecular Quadratic Zeeman Shifts

Is it really so surprising that the quadratic Zeeman shift coefficient for a particular rovibrational level should get larger as the molecular bond length increases? Why invest so much time and energy in trying to explain this quirk of molecular physics? One reason is that from a certain perspective, this quirk seems to contradict "common sense" intuition about the relationship between the properties of molecules and their constituent atoms. One might expect that as a molecule grows larger and larger, it would start to behave more and more like an unbound pair of atoms separated at infinity. Certainly the rovibrational level energy of an infinitely large molecule is simply the sum of the electronic energies of its constituent atoms. So perhaps as molecules grow larger, their quadratic Zeeman shifts also approach in some sense the sum of those of their constituent atoms.

But our data dramatically show that this is not the case. The quadratic Zeeman shift coefficient of the $^3P_1(m_j = 0)$ state is only ~ 0.179 Hz/G^2 (see calculation below), while for the weakly-bound $0_u(-1, 1, 0)$ state it is $-0.355(31)$ MHz/G^2. That represents more than a million-fold enhancement over the atomic value for two atoms which are separated by more than 400 Bohr radii, a disagreement which would diverge even more strongly if the atoms were separated to larger distances. Discovering patterns such as these helps to inform our intuition about where and when we can apply classical ideas, and when we're forced to use the full machinery of quantum mechanics to understand the results of our experiments. We'll see this pattern again when we look at how the lifetimes of subradiant states depend upon bond length in Chap. 6.

4.3.3.1 Aside: Calculation of Quadratic Shift Coefficient for $^3P_1(m = 0)$ Atoms

We can calculate the expected second order Zeeman shift once again using second order perturbation theory. The $m_j = 0$ component of the $^3P_1$ state will have zero linear Zeeman shift because $\langle ^3P_1, m_j = 0|\hat{H}_Z|^3P_1, m_j = 0\rangle = 0$, where $\hat{H}_Z = \frac{e}{2m_e}(\hat{L}_Z + g_s\hat{S}_Z)B_z$. Second order terms do not vanish, however. The expression for a second order correction to an energy level $E^{(2)}_{^3P_1}$ is:

$$E^{(2)}_{^3P_1} = \sum_{n \neq ^3P_1} \frac{|\langle ^3P_1, m_j = 0|H'|n\rangle|^2}{E_{^3P_1} - E_n} \tag{4.36}$$

We should technically sum over all excited states in the strontium atom, but since there are only two excited states within a few nanometers of $^3P_1$ (i.e., $^3P_0$ and $^3P_2$), the denominator guarantees that contributions from these two states will dominate. So we can rewrite our sum explicitly:

$$E^{(2)}_{^3P_1} \approx \sum_{m_j} \left(\frac{|\langle ^3P_1, m_j = 0|\hat{H}_Z|^3P_0, m_j = 0\rangle|^2}{E_{^3P_1} - E_{^3P_0}} + \frac{|\langle ^3P_1, m_j = 0|\hat{H}_Z|^3P_2, m_j\rangle|^2}{E_{^3P_1} - E_{^3P_2}} \right) \tag{4.37}$$

4.3 Quadratic (and Higher Order) Zeeman Shifts

In order to calculate these terms, we need to know how $\hat{H}_Z$ acts on the ^3P_j states. It's easiest if we change to the eigenbasis of the $\hat{H}_Z$ operator by writing:

$$|^3P_x, m_j\rangle \equiv |J = x, m_j\rangle_{(L=1,S=1)} = \sum_{m_l+m_s=m_j} C^{L=1,S=1,J=x}_{m_l,m_s,m_j} |L=1, m_l\rangle|S=1, m_s\rangle, \tag{4.38}$$

where $C^{L,S,J}_{m_l,m_s,m_j}$ is a Clebsch-Gordan coefficient.

Expanding all three states of importance (see, e.g., Griffiths Quantum Mechanics 2nd Edition, Section 4.4.3 for an explanation of how to read Clebsch-Gordan tables [5]), and using the convention $|J, m_j\rangle = \sum C^{L,S,J}_{m_l,m_s,m_j} |L, m_l\rangle|S, m_s\rangle$, and $\langle J, m_j| = \sum C^{L,S,J}_{m_l,m_s,m_j} \langle L, m_l|\langle S, m_s|$, we find:

- $|^3P_0, m_j = 0\rangle = |0, 0\rangle_{(L=1,S=1)} = \sqrt{\frac{1}{3}}|1, 1\rangle|1, -1\rangle - \sqrt{\frac{1}{3}}|1, 0\rangle|1, 0\rangle + \sqrt{\frac{1}{3}}|1, -1\rangle|1, 1\rangle$
- $|^3P_1, m_j = 0\rangle = |1, 0\rangle_{(L=1,S=1)} = \sqrt{\frac{1}{2}}|1, 1\rangle|1, -1\rangle - \sqrt{\frac{1}{2}}|1, -1\rangle|1, 1\rangle$
- $|^3P_2, m_j = 0\rangle = |2, 0\rangle_{(L=1,S=1)} = \sqrt{\frac{1}{6}}|1, 1\rangle|1, -1\rangle + \sqrt{\frac{2}{3}}|1, 0\rangle|1, 0\rangle + \sqrt{\frac{1}{6}}|1, -1\rangle|1, 1\rangle$

Note that of the possible $^3P_2$ terms, only $m_j = 0$ will contribute to the energy perturbation. This is because higher m_j terms will be built out of states orthogonal to $|1, 1\rangle|1, -1\rangle$ and $|1, -1\rangle|1, 1\rangle$, i.e. orthogonal to $|^3P_1\rangle$.

Acting on $^3P_0$ and $^3P_2$ with the perturbing Hamiltonian gives:

$$\hat{H}_Z|^3P_0, m_j = 0\rangle$$

$$= \frac{e}{2m_e}(\hat{L}_Z + g_s \hat{S}_Z)B_z \left(\sqrt{\frac{1}{3}}|1, 1\rangle|1, -1\rangle - \sqrt{\frac{1}{3}}|1, 0\rangle|1, 0\rangle + \sqrt{\frac{1}{3}}|1, -1\rangle|1, 1\rangle\right)$$

$$= \frac{e\hbar}{2m_e}(1 - g_e)\sqrt{\frac{1}{3}} B_z (|1, 1\rangle|1, -1\rangle - |1, -1\rangle|1, 1\rangle) \tag{4.39}$$

$$\hat{H}_Z|^3P_2, m_j = 0\rangle$$

$$= \frac{e}{2m_e}(\hat{L}_Z + g_s \hat{S}_Z)B_z \left(\sqrt{\frac{1}{6}}|1, 1\rangle|1, -1\rangle - \sqrt{\frac{2}{3}}|1, 0\rangle|1, 0\rangle + \sqrt{\frac{1}{6}}|1, -1\rangle|1, 1\rangle\right)$$

$$= \frac{e\hbar}{2m_e}(1 - g_e)\sqrt{\frac{1}{6}} B_z (|1, 1\rangle|1, -1\rangle - |1, -1\rangle|1, 1\rangle) \tag{4.40}$$

With these results in hand, we may evaluate the numerators of the terms in Eq. (4.37):

$$\langle {}^3P_1, m_j = 0|\hat{H}_Z|{}^3P_0, m_j = 0\rangle = \cdots$$

$$\cdots = \sqrt{\frac{1}{2}}\big(\langle 1,1|\langle 1,-1| - \langle 1,-1|\langle 1,1|\big)\frac{e\hbar}{2m_e}(1-g_e)\sqrt{\frac{1}{3}}B_z\big(|1,1\rangle|1,-1\rangle - |1,-1\rangle|1,1\rangle\big)$$

$$\cdots = \frac{e\hbar}{2m_e}(1-g_e)\sqrt{\frac{1}{6}}B_z\big(\langle 1,1|\langle 1,-1| - \langle 1,-1|\langle 1,1|\big)\big(|1,1\rangle|1,-1\rangle - |1,-1\rangle|1,1\rangle\big)$$

$$\cdots = \frac{e\hbar}{2m_e}(1-g_e)\sqrt{\frac{1}{6}}B_z \cdot 2 \tag{4.41}$$

$$\langle {}^3P_1, m_j = 0|\hat{H}_Z|{}^3P_2, m_j = 0\rangle = \cdots$$

$$\cdots = \sqrt{\frac{1}{2}}\big(\langle 1,1|\langle 1,-1| - \langle 1,-1|\langle 1,1|\big)\frac{e\hbar}{2m_e}(1-g_e)\sqrt{\frac{1}{6}}B_z\big(|1,1\rangle|1,-1\rangle - |1,-1\rangle|1,1\rangle\big)$$

$$\cdots = \frac{e\hbar}{2m_e}(1-g_e)\sqrt{\frac{1}{12}}B_z\big(\langle 1,1|\langle 1,-1| - \langle 1,-1|\langle 1,1|\big)\big(|1,1\rangle|1,-1\rangle - |1,-1\rangle|1,1\rangle\big)$$

$$\cdots = \frac{e\hbar}{2m_e}(1-g_e)\sqrt{\frac{1}{12}}B_z \cdot 2 \tag{4.42}$$

Finally squaring and summing gives the perturbation energy:

$$\boxed{E^{(2)}_{{}^3P_1} = \frac{1}{3}\left(\frac{e\hbar}{2m_e}\right)^2(1-g_e)^2\left(\frac{2}{E_{{}^3P_1} - E_{{}^3P_0}} + \frac{1}{E_{{}^3P_1} - E_{{}^3P_2}}\right)} \tag{4.43}$$

(Note that this result was also calculated for magnesium [4], though the cited reference leaves out intermediate steps and simply gives the result.)

The NIST handbook for strontium [7, 17] gives detunings for ${}^3P_x$ from ${}^1S_0$ in cm^{-1}. Plugging these values in gives:

$$\boxed{E^{(2)}_{{}^3P_1} = 1.19 \cdot 10^{-26}\frac{J}{T^2} = 1.79 \cdot 10^7\frac{Hz}{T^2} = 0.179\frac{Hz}{G^2}} \tag{4.44}$$

This calculated result is very close to the recently measured quadratic Zeeman shift of the ${}^3P_0$ state in ^{88}Sr [13], as might be expected.

4.3.4 Determination of Higher (up to Sixth) Order Zeeman Shifts

Stopping at linear plus quadratic shifts is a somewhat arbitrary distinction to make, since there is no reason why perturbation theory should stop there. But while the truly exact description of the Zeeman shifts will be a complicated polynomial of infinite (or close to it) order, at small magnetic fields approximating the shift as linear plus quadratic is well-justified. This is the reason why, in addition to Zeeman shift coefficients β_1 and β_2, we have also included the magnetic field ranges over which the data was taken in Tables 4.1, 4.2, 4.3, and 4.4.

That being said, it *is* interesting to ask whether we might be able to observe cases where the linear plus quadratic approximation breaks down. And it turns out, we *have* been able to observe such cases, sometimes quite dramatically. Figure 4.2 shows the shifts of certain sublevels of the $1_u(-1, J)$ states with $J = 1, 2, 3, 4$, vs magnetic field at fields as large as $\sim$50 Gauss. With the exception of the $1_u(-1, 3)$ state, these plots show data originally published by our group in 2015 [11]. Figure 4.2b in particular shows a beautiful "octopus-like" plot which clearly requires higher than second order terms for a full description. The full set of higher order terms used to describe these plots is given in Table 4.5, and (with the possible exception of the $1_u(-1, 3)$ state, which has not been studied theoretically), the fits describing this data have been shown to be in good agreement with theory calculations from a quantum chemistry model, even when they involve terms up to sixth order in magnetic field [11].

4.3.4.1 Discussion of Fit-Determined Uncertainties

The fits shown in Fig. 4.2 and summarized in Table 4.5 were produced by plotting the data using OriginLab graphing software. The uncertainties in the shift coefficients are those which are calculated by Origin's fitting routines when assigning values to the coefficients of the fitting function. The choice of fit function was made by determining how many terms were necessary to fully summarize the data. Operationally, this meant examining the fit residuals for structure, and adding higher degree terms until the fit residuals were randomly distributed about zero. This process is illustrated for the $1_u(-1, 1)$ state in Fig. 4.3.

4.4 Description of Magnetic Field Coils

Magnetic fields were applied to the molecules during spectroscopy by driving current through pairs of Helmholtz coils which surround the science chamber. As described in Chris Osborn's thesis [15], pairs of Helmholtz coils are arranged for each spatial direction, allowing the net magnetic field to be zeroed before applying

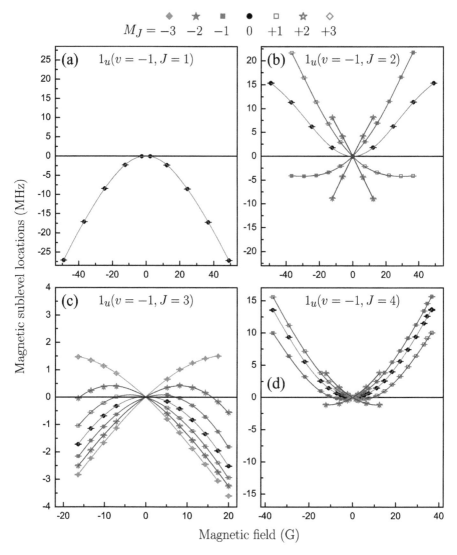

Fig. 4.2 The locations of selected sublevels of the $1_u(-1, J)$ states with (**a**) $J = 1$, (**b**) $J = 2$, (**c**) $J = 3$, and (**d**) $J = 4$ are plotted against magnetic field, showing strongly nonlinear (and even non-quadratic) shifts in many cases. The data depicting different sublevels for panels (**b**) and (**d**) were taken on different days, with slightly different calibrations for absolute laser frequency. For clarity of presentation, each pair of sublevels in these panels has been shifted so that the center of the fit coincides with the origin. Note, however, that in reality these sublevels would have different y-intercepts due to tensor light shifts, as shown in Fig. 3.13

magnetic field along the vertical (z-axis) direction. Depending on the required sensitivity of the measurement, the magnetic field gradient produced by the MOT coils can either be left on or pulsed off before molecules are created and probed.

4.4 Description of Magnetic Field Coils

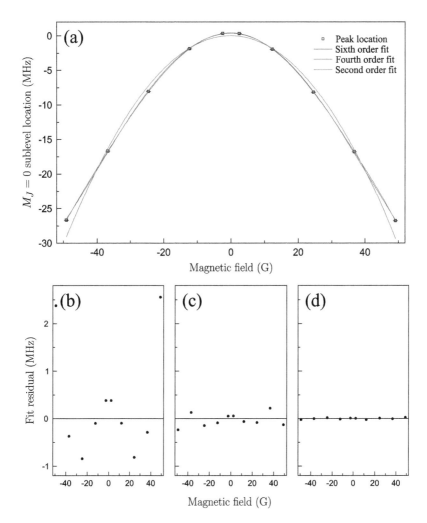

Fig. 4.3 (a) The position of the $M_J = 0$ sublevel of the $1_u(-1, 1)$ state is plotted vs magnetic field and fit with a quadratic; quadratic plus quartic; or quadratic plus quartic plus sextic fit. The residuals of the (b) quadratic and (c) quadratic plus quartic fits are clearly not randomly distributed about zero. Only the residuals of the (d) quadratic plus quartic plus sextic fit are small and random enough to have confidence that we are accurately summarizing the relevant physics

Since the atoms collect near the bottom of the MOT trapping region due to the pull of gravity, turning off the MOT coils leads to a magnetic field offset of approximately 0.9 G pointing mostly vertically. This offset is corrected for during measurements via the pulsing of a small set of compensation Helmholtz coils which are ramped on as the MOT coils are ramped off.

Whereas only small vertical fields of ∼4 Gauss could be applied during the first iteration of this experiment (i.e., from 2008–2013), we have subsequently added a new pair of coils with many more turns to achieve fields as strong as $\sim\pm60$ Gauss (where negative fields are produced simply by reversing the polarity of the Helmholtz coils), which has facilitated studies of highly nonlinear Zeeman shifts and magnetic control of transition strengths (see Chap. 5). The limiting factor for achieving large magnetic fields is the resistance of the Helmholtz coils combined with the finite power which can be supplied by our Delta Elektronika ES-030-5 power supplies. When using the coils to produce fields near their maximum, heating of the coils can cause their resistance to increase with time. If care is not taken, this increasing resistance can cause the power load to exceed the maximum power output attainable by the Delta Elektronika ES-030-5 supply, leading to apparently non-linear effects. This potential problem can be avoided by a combination of operating at small currents and only pulsing the coils for short amounts of time, so that the coils do not have time to resistively heat. (Note that the duty cycle is also important, so that the coils have time to air-cool between experimental shots.)

4.4.1 Calibration to the $^1S_0 + {}^3P_1$ Intercombination Line

Because ^{88}Sr carries zero net nuclear spin, the Zeeman shifts of its energy levels are entirely determined by magnetic field interactions with the magnetic moment produced by *electronic* angular momentum. For a ^{88}Sr atom in the singlet $^1S_0$ electronic ground state, the paired electrons in the valence shell carry no net angular momentum, and thus possess no magnetic moment. Therefore ^{88}Sr atoms in the electronic ground state have zero first-order Zeeman shift. If the atom is in an electronically excited ^3P state, however, then the paired electrons form a triplet state with orbital electronic angular momentum $l = 1$ and total spin $s = 1$. These two forms of electronic angular momentum can combine to form a total angular momentum $j = |l-s|, |l-s|+1, \ldots |l+s|$, where j can take on the values 0, 1, or 2, with projection m_j along of the quantization axis taking on the values $m_j = -j, -j+1, \ldots, j$.

We choose to calibrate our coils by examining the magnetic field dependence of the $m = \pm 1$ components of the atomic intercombination line transition ($^1S_0 \rightarrow {}^3P_1$), whose frequency shift Δf can be summarized with the following equation:

$$\Delta f = g_A \mu_B m_j B + O(B^2), \tag{4.45}$$

where g_A is the atomic Lande g-factor, μ_B is the Bohr magneton, B is the applied magnetic field, and terms of order B^2 or higher can be neglected at our magnetic field strengths (see Sect. 4.3.3.1). By plotting the measured frequency shift of this transition vs the current supplied to the coils, we can compare to the expected frequency shift vs magnetic field (Eq. (4.5)) to extract a conversion from Amperes (controlled by a DAQ-supplied Voltage) to Gauss. Figure 4.4 shows data used for the calibration of our highest-field coils.

4.4 Description of Magnetic Field Coils

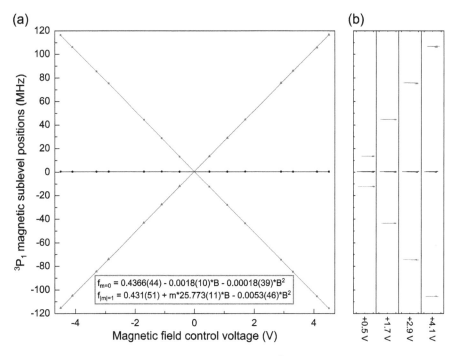

Fig. 4.4 (a) The locations of the magnetic sublevels of the $^3P_1$ atomic state (corrected for cavity drift) are plotted vs magnetic field, using our maximum field-producing coil configuration. (b) Sample spectra at several positive applied voltages. Lineshapes are produced by heating out of the lattice, and are $\sim$100 kHz broad

4.4.1.1 Spectroscopy on the $^1S_0 \rightarrow {}^3P_1$ Intercombination Line

Because the $^1S_0 \rightarrow {}^3P_1$ transition is closed, interrogating the magnetic sublevels of the excited state is slightly nontrivial. To observe losses due to the spectroscopy laser, we interrogated atoms in the 1D lattice with a long-duration (several hundred ms) and low power (few-nW) probe laser, and monitored atom losses due to heating out of the trap. This means that the data from Fig. 4.4 consisted of the locations of the blue lattice sideband, rather than carrier transitions. However, since the same lattice power was used for all measurements, results derived in this manner should be equivalent to results derived from carrier transitions.

4.4.2 Quantized Output from NI PXI-6713 Card

The currents supplied by the Delta Elektronika supplies are controlled by a programmable input voltage supplied by a NI PXI-6713 card. This card can produce voltages ranging from -10 to 10 V, but with 12 bit resolution, resulting in a small (but measurable) step size of $20 \text{ V}/2^{12} = 4.88$ mV.

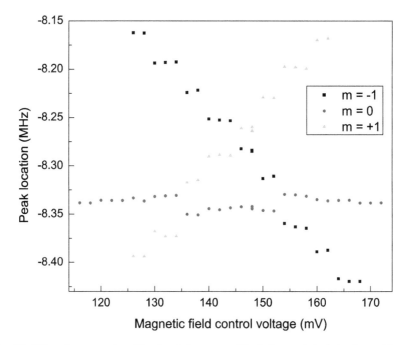

Fig. 4.5 When the magnetic sublevels of the narrow $X(-1, 0) \to 1_g(-1, 1, m)$ transition are interrogated at small magnetic fields, the peak positions appear to "jump" discretely as a result of the quantized voltage output from our PXI card. This data was recorded with the same set of field coils as Fig. 4.4, and is the unaveraged version of Fig. 3.13a

This small quantization of the output voltage results in quantized values for the magnetic field. For the largest field coil configuration, the magnetic field will be stepped by $(0.00488\,\text{V}) \times 12.275\,\text{G/V} = \sim 60\,\text{mG}$. For our smallest-field coil configuration, the step size would be $(0.00488\,\text{V}) \times 0.76073\,\text{G/V} = \sim 3.7\,\text{mG}$. Both of these step sizes could be reduced in the future by purchasing a PXI card with a higher bit resolution. Figure 4.5 shows measurements of the positions of the three magnetic sublevels of the $1_g(-1, 1)$ state vs different control voltages, showing discrete changes in the peak positions for different voltages clearly indicative of the fact that multiple values of input voltage result in the same magnetic field.

References

1. Borkowski, M., Morzyński, P., Ciuryło, R., Julienne, P., Yan, M., DeSalvo, B., Killian, T.: Mass scaling and nonadiabatic effects in photoassociation spectroscopy of ultracold strontium atoms. Phys. Rev. A **90**(3), 032713 (2014)
2. Brown, J., Carrington, A.: Rotational Spectroscopy of Diatomic Molecules. Cambridge University Press, Cambridge (2003)

3. Carrington, A., Leach, C., Marr, A., Shaw, A., Viant, M., Hutson, J., Law, M.: Microwave spectroscopy and interaction potential of the long-range He...Ar+ ion. J. Chem. Phys. **102**(6), 2379–2403 (1995)
4. Friebe, J., Pape, A., Riedmann, M., Moldenhauer, K., Mehlstäubler, T., Rehbein, N., Lisdat, C., Rasel, E., Ertmer, W., Schnatz, H., Burghard, L., Gesine, G.: Absolute frequency measurement of the magnesium intercombination transition $^1S_0 \rightarrow {}^3P_1$. Phys. Rev. A **78**(3), 033830 (2008)
5. Griffiths, D.: Introduction to Quantum Mechanics, 2nd edn. Pearson Prentice Hall, Upper Saddle River (2005)
6. Hanneke, D., Fogwell, S., Gabrielse, G.: New measurement of the electron magnetic moment and the fine structure constant. Phys. Rev. Lett. **100**(12), 120801 (2008)
7. Kramida, A., Ralchenko, Y., Reader, J.: NIST atomic spectra database (ver. 5.1). National Institute of Standards and Technology, Gaithersburg, MD (2013)
8. LeRoy, R., Bernstein, R.: Dissociation energy and long-range potential of diatomic molecules from vibrational spacings of higher levels. J. Chem. Phys. **52**(8), 3869–3879 (1970)
9. McDonald, M.: High precision optical spectroscopy and quantum state selected photodissociation of ultracold $^{88}Sr_2$ molecules in an optical lattice. Ph.D. Thesis, Columbia University in the City of New York (2016)
10. McGuyer, B., Osborn, C., McDonald, M., Reinaudi, G., Skomorowski, W., Moszynski, R., Zelevinsky, T.: Nonadiabatic effects in ultracold molecules via anomalous linear and quadratic Zeeman shifts. Phys. Rev. Lett. **111**(24), 243003 (2013)
11. McGuyer, B., McDonald, M., Iwata, G., Skomorowski, W., Moszynski, R., Zelevinsky, T.: Control of optical transitions with magnetic fields in weakly bound molecules. Phys. Rev. Lett. **115**(5), 053001 (2015)
12. McGuyer, B., McDonald, M., Iwata, G., Tarallo, M., Skomorowski, W., Moszynski, R., Zelevinsky, T.: Precise study of asymptotic physics with subradiant ultracold molecules. Nat. Phys. **11**(1), 32–36 (2015)
13. Morzyński, P., Bober, M., Bartoszek-Bober, D., Nawrocki, J., Krehlik, P., Śliwczyński, Ł., Lipiński, M., Masłowski, P., Cygan, A., Dunst, P., Garus, M., Lisak, D., Zachorowski, J., Gawlik, W., Radzewicz, C., Ciuryło, R., Zawada, M.: Absolute measurement of the $^1S_0-{}^3P_0$ clock transition in neutral ^{88}Sr over the 330 km-long stabilized fibre optic link. Sci. Rep. **5**, 17495 (2015)
14. Odom, B., Hanneke, D., D'Urso, B., Gabrielse, G.: New measurement of the electron magnetic moment using a one-electron quantum cyclotron. Phys. Rev. Lett. **97**(3), 030801 (2006)
15. Osborn, C.: The physics of ultracold Sr_2 molecules: Optical production and precision measurement. Ph.D. Thesis (2014)
16. Porsev, S., Safronova, M., Clark, C.W.: Relativistic calculations of C_6 and C_8 coefficients for strontium dimers. Phys. Rev. A **90**(5), 052715 (2014)
17. Sansonetti, J., Nave, G.: Wavelengths, transition probaAbilities, and energy levels for the spectrum of neutral strontium (Sr I). J. Phys. Chem. Ref. Data **39**(3) 033103-1 (2010)
18. Skomorowski, W., Pawłowski, F., Koch, C., Moszynski, R.: Rovibrational dynamics of the strontium molecule in the $A^1\Sigma_u^+$, $c^3\Pi_u$, and $a^3\Sigma_u^+$ manifold from state-of-the-art ab initio calculations. J. Chem. Phys. **136**(19), 194306 (2012)
19. Takasu, Y., Saito, Y., Takahashi, Y., Borkowski, M., Ciuryło, R., Julienne, P.: Controlled production of subradiant states of a diatomic molecule in an optical lattice. Phys. Rev. Lett. **108**(17), 173002 (2012)
20. Woodgate, G.: Elementary Atomic Structure. Oxford University Press, Oxford (1980)
21. Zeeman, P.: XXXII. On the influence of magnetism on the nature of the light emitted by a substance. Lond. Edinb. Dublin Philos. Mag. J. Sci. **43**, 262, 226–239 (1897)
22. Zelevinsky, T., Boyd, M., Ludlow, A., Ido, T., Ye, J., Ciuryło, R., Naidon, P., Julienne, P.: Narrow line photoassociation in an optical lattice. Phys. Rev. Lett. **96**(20), 203201 (2006)

Chapter 5
Magnetic Control of Transition Strengths

5.1 Introduction (Defining "Transition Strength")

So far we've talked about discrete rovibrational levels in ^{88}Sr$_2$ in isolation, and described how these levels can be characterized through precise measurements of their binding energies and Zeeman shifts. Both of these quantities are interesting because they reflect something about the larger scale structure of the molecule, each forming part of a "fingerprint" giving every level a subtly different flavor. The flavor of each level is inextricably related to the flavor of every other level through perturbation theory, since small fields or terms within the molecular Hamiltonian itself can cause interactions perturbing the properties of nearby levels. But while such large-scale characterization is interesting from a general perspective, we have another tool we can use to make targeted, controlled studies of the interactions between particular pairs of levels.

By characterizing the *transition strength* of a pair of levels, we evaluate the magnitude of a single matrix element $\langle 1|\hat{H}'|2\rangle$ connecting two states $|1\rangle$ and $|2\rangle$, rather than a quantity depending on the result of a sum over many different levels connected by the perturbing operator $\hat{H}'$. For a laser spectroscopy experiment such as ours, the perturbing operator describes the interaction of the laser fields with the bound states in question. The coarse behavior of these strengths is often summarized by *selection rules*, which approximate the strength of a transition as 0 or 1 depending on the initial and final quantum numbers (J, M_J, Ω, etc.) of the states being probed. But as we've seen already, these quantum numbers are sometimes only approximate. Moreover, they can be strongly modified when subjected to large external fields.

In order to make quantitative statements about these effects, we need a rigorous way to define the transition strength. A natural place to begin would be to start with Einstein's A and B coefficients, which represented the first successful attempt to characterize the strength of the interactions between light and matter [1]. Specifically, however we define our transition strength should be proportional to the induced absorption coefficient B_{12}^ω.

5.2 Three Ways to Measure

We've studied three different experimentally observable quantities which can be related to B_{12}^ω. Our choice of which to use in a particular situation depends upon the experimental accessibility of the transition under investigation.

5.2.1 Normalized Area Under a Lorentzian

A typical molecular spectroscopy experiment in our lab has the following structure. First, a sample of molecules is prepared in some initial state. Next, a spectroscopy laser pulse is applied which is resonant with a transition to some final state, transferring some amount of population. Finally, the amount of population remaining in the initial state after spectroscopy is measured. If the transition is open (i.e., if spontaneous decay from final to initial state is minimal), and if the duration of the probe pulse is much longer than the lifetime of the final state, then we can describe the observed initial state population $N(t)$ with the following rate equation:

$$\frac{d}{dt}N(t) = -\Gamma(\delta)N(t) \tag{5.1}$$

where $\Gamma(\delta)$ is a function of the laser detuning from resonance δ. The solution to this equation gives the population remaining in the initial state after spectroscopy with a probe pulse of duration τ:

$$N(\tau) = N(0)e^{-\Gamma(\delta)\tau}. \tag{5.2}$$

To characterize the total transition strength, we define the following experimental quantity Q:

$$Q = \frac{1}{P}\int \Gamma(\delta)d\delta, \tag{5.3}$$

where P is the laser power used to drive the transition. Because the signal we measure is proportional not to $\Gamma(\delta)$, but rather to the initial state population $N(\tau)$, we can rewrite Eq. (5.3) as:

5.2 Three Ways to Measure

$$Q = \frac{1}{\tau P} \int \ln\left[\frac{S(\delta)}{S(\infty)}\right] d\delta \equiv \frac{-A}{\tau P}, \quad (5.4)$$

where the signal $S(\delta) \propto N(\tau) = N(0)e^{-\Gamma(\delta)\tau}$ has been written to emphasize that the total scaling for the molecule number is unimportant, and A is defined as the total area underneath the natural log of the total signal curve.

We have defined Q in this way because it is a relatively simple quantity to measure, requiring knowledge of only three quantities (probe power, pulse time, and area under a spectroscopic curve). For Q to be a "good" description of the transition strength, however, it should be proportional to the induced absorption coefficient B_{12}^ω. We can show that this is in fact the case in the following way.

Following Hilborn [3], we recognize that $\Gamma(\delta)$ is an induced absorption rate per molecule

$$\Gamma(\delta) = \frac{W_{12}^i}{N_1}, \quad (5.5)$$

where W_{12}^i is the total rate of induced absorption and N_1 is the population of the initial state. We can rewrite W_{12}^i using Eqs. (5.17)–(5.19) of Hilborn:

$$W_{12}^i = \int w_{12}^i(\omega)d\omega = N_1 \int b_{12}(\omega)\rho(\omega)d\omega = N_1 B_{12}^\omega \int g(\omega)\rho(\omega)d\omega, \quad (5.6)$$

where $\rho(\omega)$ is the energy density per angular frequency at ω, $b_{12}(\omega) = B_{12}^\omega g(\omega)$, and $g(\omega)$ is a normalized transition lineshape function satisfying $\int g(\omega)d\omega = 1$.

For a nearly monochromatic directional light beam (e.g., a laser) we can relate the total irradiance I (i.e., the total power per unit area received by the molecule) to the energy density per angular frequency $\rho(\omega)$ with the following formula:

$$I = \int c\rho(\omega)d\omega = \int i(\omega - 2\pi\delta)d\omega, \quad (5.7)$$

where $i(\omega - 2\pi\delta)$ is a function describing the lineshape of a laser with peak intensity at frequency δ. (Note that in the limit of a very narrow linewidth laser, $i(\omega - 2\pi\delta) \to I\delta(\omega - 2\pi\delta)$, where $\delta(\omega - 2\pi\delta)$ is the Dirac delta function.) Substituting $\rho(\omega) = \frac{1}{c}i(\omega - 2\pi\delta)$ into Eq. (5.6), we can rewrite $\Gamma(\delta)$ as:

$$\Gamma(\delta) = \frac{B_{12}^\omega}{c} \int g(\omega)i(\omega - 2\pi\delta)d\omega \quad (5.8)$$

Plugging this result into our definition of Q from Eq. (5.3) gives the following expression:

$$Q = \frac{B_{12}^\omega}{cP} \int d\delta \int d\omega \cdot g(\omega)i(\omega - 2\pi\delta), \quad (5.9)$$

By Fubini's theorem, we can reverse the order of integration in the above double integral. Since the limits of integration are over all frequencies, we can use $\int i(\omega - 2\pi\delta)d\delta = \frac{1}{2\pi}I$ and $\int g(\omega)d\omega = 1$ to reach our final result:

$$Q = \frac{IB_{12}^{\omega}}{2\pi cP} = \frac{B_{12}^{\omega}}{c\pi^2 w_0^2}, \tag{5.10}$$

where for the second equality I have made use of the result that the maximal irradiance I for a Gaussian laser beam is related to its maximal power P and its waist w_0 by $I = \frac{c\epsilon_0|\vec{E}|^2}{2} = \frac{2P}{\pi w_0^2}$.

Note that relationship between Q and B_{12}^{ω} depends upon the value of the probe waist w_0. This means that so long as the same probe laser optics are used to make Q-measurements (i.e., so long as w_0 remains unchanged), different measurements can be compared against one another. For relative transition strength measurements presented in this thesis, we were careful to make measurements as quickly as possible to minimize the risk of the effective waist changing.

5.2.1.1 Accounting for Degeneracy in the Initial and Final States (or, the Consequences of "m-Mixing")

The above derivation of the relationship between Q and B_{12}^{ω} assumes no degeneracy in the initial state, implying that the population being measured is exactly the population being probed. While this is true for experiments starting from $J = 0$, this assumption can be violated for experiments starting from $J = 2$ due to *mixed quantization*.

The quantization axis will be primarily defined by the orientation of whatever field causes the largest splitting among sublevels. The electronic excited states of ^{88}Sr$_2$ possess strong Zeeman shifts due to the angular momentum projection of the $^3P_1$ atom, meaning that the quantization axis will be defined by the magnetic field orientation. Ground state ^{88}Sr$_2$, however, is very nearly non-magnetic: whereas linear Zeeman shifts in the excited state will be on the order of the Bohr magneton ($\sim$1.4 MHz/G), in the ground state they will be of the order of the nuclear magneton ($\sim$760 Hz/G)—a factor of nearly 2000 smaller, and potentially measurable only at our lab's highest achievable magnetic fields (though so far not unambiguously observed). If no other sublevel-perturbing fields were present, then the ground state sublevels would simply be undefined, and transitions would begin from an incoherent mixture of many possible initial sublevels. In our case, however, our molecules are probed in a linearly-polarized 1D optical lattice, which induces small (tens of kHz) tensor light shifts among different magnetic sublevels. Since these tensor light shifts are much larger than the magnetic Zeeman shifts in this case, *the quantization axis for ground state ^{88}Sr$_2$ is defined by the polarization orientation of the lattice*.

5.2 Three Ways to Measure

We can choose to probe molecules with the lattice polarization axis either parallel or perpendicular to the magnetic field. If the two axes are perpendicular (as they would be when operating in the "magic wavelength" condition at ∼914 nm for the $^1S_0+{}^3P_1$ transition [4]), then the same basis cannot be used to describe both initial and final states. In other words, the Hamiltonian for the final state will have off-diagonal elements when written in the basis of the initial state [6]. We call this phenomenon *m-mixing*.

The consequences of this effect are dramatically illustrated in Fig. 5.1. The plots show interrogation of the $X(-2, J = 2)$ state, which possesses five magnetic sublevels (Fig. 5.1a). In panels (b)–(d), a depletion laser is resonantly applied to preferentially deplete a single sublevel, after which the magnetic sublevel distribution in the initial state is probed by sweeping a laser across a weakly-bound recovery transition. The surprising result is that *depleting a magnetic sublevel m also (apparently) depletes sublevels $m\pm2$, $m\pm4$, and so on*. This is a wonderfully bizarre result that defies classical intuition. We cannot think of the magnetic sublevels as "good" quantum labels, but in this case must instead consider that the basis functions of the initial state are best represented as superpositions of those of the final state [6]. Panels (e)–(g) depict a similar experiment from a complementary perspective. See the figure caption for details.

While this all sounds horribly complicated and undesirable, there are in fact cases for which we can use *m*-mixing to our advantage. For example, since we produce our ground-state molecules via one-photon photoassociation followed by spontaneous decay, then when the lattice polarization is parallel to the magnetic field we are at the mercy of the selection rule $\Delta m = 0, \pm 1$ to determine the distribution of magnetic sublevels in the ground state (e.g., to produce $m = -2$ we must photoassociate to the $m' = 1$ sublevel). Furthermore, inconvenient branching ratios in this case might ensure that the population of the target sublevel is only a small fraction of the total population. However, if we choose for our lattice to be perpendicular to the magnetic field, then transitions starting from $m = -2$ can be observed simply by populating $m = 0$.

The Effect of *m*-Mixing on Transition Strength Measurements The fact that a single probe laser can simultaneously deplete several magnetic sublevels implies that we should think carefully about what the "strength" we measure means, since m is no longer a good quantum number. In order to correct for this effect in cases where it is applicable, we in practice multiply the measured quantity Q_mix (obtained in the way described in the previous section) by a correction factor R to obtain the "true" transition strength Q according to:

$$Q(m_1, m') = R(m_1, m_2) Q_\text{mix}(m_1, m', m_2), \qquad (5.11)$$

where m_1 is the magnetic sublevel of the *probed* initial state, m_2 is the magnetic sublevel of the *detected* state, and m' is the magnetic sublevel of the *probed* final state (where for π-transitions, $m' = m$). For information on how to calculate this correction factor, see the supplement of reference [6].

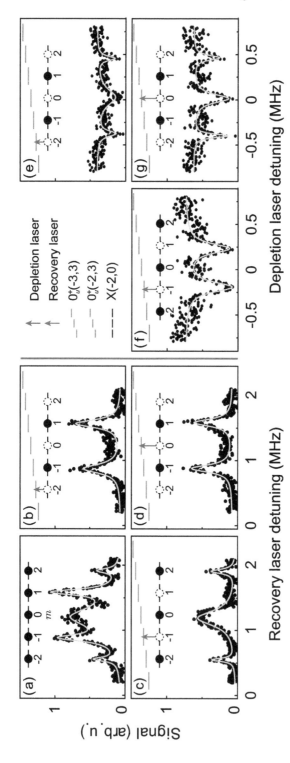

Fig. 5.1 A sample of $X(-2,2)$ molecules is produced via photoassociation to the $0_u^+(-5,1,0)$ state, with magnetic field oriented vertically and lattice polarized horizontally. (**a**) The molecules are immediately recovered by sweeping a 100 μs duration "Recovery laser" pulse across the transition from $X(-2,2)$ to $0_u^+(-2,3,m')$, showing all five magnetic sublevels present. (**b**)–(**g**) Before recovery, a 1 ms duration "Depletion laser" pulse is applied which is resonant with a transition from $X(-2,2,m)$ to $0_u^+(-3,3,m')$. (**b**)–(**d**) show spectra resulting from parking the depletion laser on resonance and then sweeping the recovery laser, while (**e**)–(**g**) show spectra resulting from sweeping the depletion laser and then parking the recovery laser on resonance. Adapted partially from [6]

5.2 Three Ways to Measure

5.2.1.2 Effect of Power Saturation

Note that the previous derivation is only strictly correct in the limit of small probe powers. At large probe powers, i.e. powers comparable to the saturation power P_{sat}, power broadening can significantly alter the measured value of the transition strength. For example, if the transition can be approximated at large probe powers as a two-level system described by a Lorentzian lineshape, then the form of $\Gamma(\delta)$ from Eq. (5.1) will be [9]

$$\Gamma(\delta) = \frac{\gamma^2(1+s_0)^{\frac{1}{2}}}{4} \cdot \left[\frac{s_0}{1+s_0}\right] \cdot \frac{\frac{\gamma}{2}(1+s_0)^{\frac{1}{2}}}{\left[\frac{1}{2}\gamma(1+s_0)^{\frac{1}{2}}\right]^2 + \delta^2}, \quad (5.12)$$

where γ is the natural linewidth of the molecular transition (technically the decay rate from final to initial state in this approximation) and the saturation parameter $s_0 = (P/P_{\text{sat}})$ (assuming a uniform probe intensity across the cloud). Plugging this into our definition of Q from Eq. (5.3) and integrating over all frequencies yields the following:

$$Q = \frac{\gamma^2(1+s_0)^{\frac{1}{2}}}{4P} \cdot \left[\frac{s_0}{1+s_0}\right] \int_{-\infty}^{+\infty} \frac{\frac{\gamma}{2}(1+s_0)^{\frac{1}{2}}}{\left[\frac{1}{2}\gamma(1+s_0)^{\frac{1}{2}}\right]^2 + \delta^2} d\delta = \frac{\gamma^2 \pi}{4P} \cdot \left[\frac{s_0}{\sqrt{1+s_0}}\right]$$

$$(5.13)$$

Since $s_0 \propto P$, our definition of Q is independent of the probe power used to perform the measurement in the limit of low power (as it should be!). At higher powers, however, where the denominator $\sqrt{1+s_0}$ differs significantly from 1, the measured value of Q becomes noticeably smaller than the "true" value of the transition strength.

We have in fact observed that in regimes where power broadening is obvious, our measured values of transition strength *decrease* (as expected) with increasing probe power. See, e.g., Fig. 6.1b. Because of this, we have been careful to make spectroscopic transition strength measurements at very low probe powers, where power broadening is negligible.

5.2.2 Rabi Oscillations

Consider a situation in which a probe laser is tuned very close to resonance with states $|1\rangle$ and $|2\rangle$, and is strong enough to drive transitions at a rate much faster than the natural decay rate Γ of state $|2\rangle$. The Hamiltonian $\hat{H}(t)$ describing such a system can be divided into two parts:

$$\hat{H}(t) = \hat{H}_0 + \hat{H}_I \cos(\omega t), \quad (5.14)$$

where $\hat{H}_0|n\rangle = E_n|n\rangle$, and $\hat{H}_I \cos(\omega t)$ is the time-dependent perturbation induced by the laser [2].

Neglecting for now decay to other states, we can write the wavefunction $|\Psi(t)\rangle$ for the system as

$$|\Psi(t)\rangle = c_1(t)|1\rangle + c_2(t)|2\rangle, \tag{5.15}$$

where the variables $c_1(t)$ and $c_2(t)$ represent the probability amplitudes for finding the molecule in state $|1\rangle$ or $|2\rangle$, respectively. It's relatively straightforward [2] to show that by applying the time-dependent Schroedinger equation to Eq. (5.15), we find the following pair of differential equations describing the population evolution between states $|1\rangle$ and $|2\rangle$:

$$i\dot{c}_1 = \Omega_{12} \cos(\omega t) e^{-i\omega_0 t} c_2$$
$$i\dot{c}_2 = \Omega_{12}^* \cos(\omega t) e^{i\omega_0 t} c_1, \tag{5.16}$$

where $\omega_0 = \frac{1}{\hbar}(E_2 - E_1)$, and the molecular Rabi frequency Ω_{12} is given by

$$\Omega_{12} = \frac{1}{\hbar} \langle 1|\hat{H}_I|2\rangle, \tag{5.17}$$

where for E1, M1, and E2 transitions the interaction Hamiltonian has the following forms:

- $\hat{H}_{I,E1} = -\hat{d} \cdot \vec{E}_0$, where $\hat{d}$ is the electric dipole moment operator and $\vec{E}_0$ is the electric field
- $\hat{H}_{I,M1} = -\hat{\mu} \cdot \vec{B}_0$, where $\hat{\mu}$ is the magnetic dipole moment operator and $\vec{B}_0$ is the magnetic field
- $\hat{H}_{I,E2} = -(1/6)\hat{Q}_{ij}\nabla_i E_j$, where $\hat{Q}_{ij}$ is the electric quadrupole moment operator and E_j is the j-th component of the electric field

To solve this pair of differential equations, it's common to expand $\cos(\omega t) = \frac{1}{2}(e^{i(\omega-\omega_0)} + e^{i(\omega+\omega_0)})$ and then to make the *rotating wave approximation*, which assumes that when $\omega \approx \omega_0$ the evolution of the system will be dominated by slowly oscillating terms, and therefore that terms proportional to $e^{i(\omega+\omega_0)}$ can be thrown out. If we make this approximation, we get the following equations:

$$i\dot{c}_1 = c_2 e^{i\delta t} \frac{\Omega_{12}}{2}$$
$$i\dot{c}_2 = c_1 e^{-i\delta t} \frac{\Omega_{12}^*}{2}, \tag{5.18}$$

where $\delta \equiv \omega - \omega_0$.

At resonance (i.e., $\delta = 0$), and assuming the population starts entirely in state $|1\rangle$, the solution to Eqs. (5.18) describes sinusoidally oscillating populations of initial and final states:

5.2 Three Ways to Measure

$$|c_1(t)|^2 = \cos^2\left(\frac{\Omega_{12}t}{2}\right)$$
$$|c_2(t)|^2 = \sin^2\left(\frac{\Omega_{12}t}{2}\right),$$
(5.19)

The variable Ω_{12} is called the *Rabi frequency*, and represents the frequency at which population oscillates between states $|1\rangle$ and $|2\rangle$ due to the presence of the probe laser.

5.2.2.1 Comment on Coherence Time and Natural Linewidth

In our lab we have achieved record molecule-light coherence times [8]. Despite this, the Rabi oscillations we observe are still not very well described by the solutions to Eqs. (5.18), since the coherence times we achieve are on the order of a few hundred microseconds, limited mainly by the lifetimes of our longest-lived subradiant states (see Chap. 6) and comparable to the durations of our probe pulses.

We can achieve a better fit to our data by building in a mechanism for spontaneous decay. One way to do so is to modify Eqs. (5.18) in the following way:

$$i\dot{c}_1 = c_2 e^{i\delta t}\frac{\Omega_{12}}{2}$$
$$i\dot{c}_2 = c_1 e^{-i\delta t}\frac{\Omega_{12}^*}{2} - \frac{i\Gamma}{2}c_2$$
(5.20)

Note that in the limit of no laser coupling between states $|1\rangle$ and $|2\rangle$, a population starting initially entirely in state $|2\rangle$ would evolve according to

$$|c_2(t)|^2 = e^{-\Gamma t},$$
(5.21)

which is exactly what we'd expect for spontaneous decay.

The general solution to Eqs. (5.20) for $\delta \neq 0$ is complicated. But if we assume that our probe laser is on resonance before solving, we get the following result:

$$|c_1(t)|^2 = \begin{cases} e^{-\frac{\Gamma}{2}t}\cos^2\left[\left(\Omega_{12}^2 - \frac{\Gamma^2}{4}\right)^{\frac{1}{2}}t\right], & \text{when } \Omega_{12} > \frac{\Gamma}{2} \\ e^{[-\frac{\Gamma}{2} - (\frac{\Gamma^2}{4} - \Omega_{12}^2)^{\frac{1}{2}}]t}, & \text{when } \Omega_{12} < \frac{\Gamma}{2} \end{cases}$$
(5.22)

The full solution can be well approximated by adding small offsets for y-intercept and phase to Eq. (5.22). This has been done in fits to Rabi oscillation data for subradiant states described in a later chapter. The Rabi frequency Ω_{12} is then extracted from the fit and plotted against power in order to determine the transition strength.

5.2.2.2 Relating Transition Strength Q to Rabi Frequency Ω_{12}

We can once again turn to Hilborn [3] to determine how our measurement of Rabi frequency will relate to transition strength. In that reference's Table 1, we find the following relationship:

$$\mu_{12}^2 = \frac{(\hbar\Omega_{12})^2}{E^2} = 3\frac{g_1}{g_2}\frac{\epsilon_0 \hbar^2}{\pi}B_{12}^{\omega} \tag{5.23}$$

From Eq. (5.10) we know that $Q = \frac{IB_{12}^{\omega}}{2\pi cP}$. Using once again the relationship between electric field and irradiance, $I = \frac{c\epsilon_0|\vec{E}|^2}{2}$, we find [6, 8]:

$$Q = \frac{g_2}{g_1} \cdot \frac{\Omega_{12}^2}{12P}, \tag{5.24}$$

where g_1 and g_2 are the magnetic sublevel degeneracies of the initial and final states, respectively. For transitions from $J = 0 \rightarrow J' = 1$ (which are the only transitions described in this thesis for which Rabi flopping measurements have been made), $g_2/g_1 = 3$.

5.2.3 Autler-Townes Splitting (Two-Photon Spectroscopy)

Determining transition strength via a measurement of the *Rabi-flopping frequency* (Sect. 5.2.2) is only possible when the transitions can be probed on timescales shorter than the lifetime of the excited state. In our experiment, we've found that this is possible for only a few *subradiant* states within the 1_g potential, while for the more easily-accessible *superradiant* 1_u states (with $\sim 10\,\mu s$ lifetimes), Rabi oscillations are damped out too quickly to allow for accurate characterization.

Spectroscopic determination of a transition strength via measurement of a normalized area under a lineshape (Sect. 5.2.1) is less restrictive, requiring only the possibility of producing a stable population in the initial state from which the transition will be excited, and is in fact better suited to situations where probing is incoherent and can be described by a rate equation. However, it is not easy to produce such populations in arbitrary rovibrational states. We take advantage of favorable branching ratios to produce sizable samples of molecules in the $X(v = -1; J = 0, 2)$ and $X(v = -2; J = 0, 2)$ ground states with only a single laser, but to produce populations in more deeply-bound ground states would require more lasers (possibly phase-locked to one another via a frequency-comb), which, as of the writing of this thesis, has not been conclusively demonstrated by our lab.

In cases for which the previous two techniques fail, a third option is available: measurement of a transition's Rabi frequency via two-photon *Autler-Townes spectroscopy*, whereby a photoassociation spectrum is split into a doublet upon simultaneous application of a second probe laser tuned to resonance with the transition to be studied. This technique has been used by ours and other experimental groups to determine ground state binding energies to few-hundred kHz precision [5, 11], and was described in detail in Chap. 3. The salient details relating to determination of transition strength are described below.

5.2.3.1 Theory

From Chap. 3, we know that the photoassociation spectrum describing an Autler-Townes doublet can be described by a frequency-dependent PA rate K given by Eq. (3.10) (reproduced here):

$$K(\epsilon, \delta_1) = C \frac{(\epsilon/h - \Delta_2)^2}{[(\epsilon/h - \Delta_+)(\epsilon/h - \Delta_-)]^2 + (\gamma/2)^2(\epsilon/h - \Delta_2)^2}, \quad (5.25)$$

with the following variable definitions:

- $\Delta_\pm = \frac{1}{2}(\Delta_1 + \Delta_2) \pm \frac{1}{2}\sqrt{(\Delta_1 - \Delta_2)^2 + 4\hbar^2\Omega_{12}^2}$
- $\Delta_1 = -(\delta_1 - \delta_{1c})$
- $\Delta_2 = \delta_{2c} - (\delta_1 - \delta_{1c})$

The variable "Ω_{12}" is just the molecular Rabi frequency given by Eq. (5.17). Therefore we can follow the same technique as we did in the previous section, namely measure the transition strength $Q = \frac{\Omega_{12}^2}{4P}$, where P is the power of the bound–bound laser used to split the doublet [10].

This method is more generally applicable than the Rabi oscillation method because it can be applied to even very broad transitions. However, it is also much more time-intensive than the previous two techniques. Whereas spectroscopic and Rabi oscillation measurements require only ~50 points (varying either laser frequency or duration) to extract a transition strength, Autler-Townes spectroscopy requires several spectroscopic traces with the bound–bound laser at various detunings to confidently extract a Rabi frequency (see Fig. 3.3).

5.2.3.2 Results

Table 5.1 shows measurements of Ω_{12}^2/P for transitions between weakly-bound levels. The particular levels shown in this table were chosen because they were predicted to have the largest transition strengths, and therefore held the most promise for producing large samples of ground state molecules via one-photon photoassociation. Values for the Rabi frequency Ω_{12} were obtained by fitting Eq. (3.10) to two-photon PA spectra. A detailed description of this fitting process, as well an illustration of the data used to create the $X(-3, 0) \rightarrow 0_u(-6, 1, 0)$ entry, is given in Sect. 3.2.3.1.

Table 5.1 This table shows measurements of Ω_{12}^2/P via two-photon photoassociation (Autler-Townes spectroscopy) for transitions between weakly–bound levels thought to be relevant to ground-state molecule production (i.e., those predicted to have the largest strengths)

$X\,^1\Sigma_g^+$		$(1)0_u^+$					$(1)1_u$
		v':	-3	-4	-5	-6	-3
v	E_b	E_b':	222	1084	3463	8429	8200
-1	137		0.66(8)	0.53(2)			
			0.606	0.545			
-2	1400			0.158(6)	1.00(8)		
				0.144	1		
-3	5111					1.88(9)	0.46(3)
						1.651	0.553

Because of the difficulty in measuring laser probe waist, *relative* strengths are shown, normalized to the $X(-2,0) \to 0_u(-5,1,0)$ transition. Uncertainties are indicated on experimental points, and ab initio theoretical values [7, 12] are shown beneath the measured values. The strengths of transitions to the three more weakly-bound levels were measured within a few hours of one another, while more deeply-bound points were measured the next day. Adapted from [7]

5.3 Enabling "Forbidden" Transitions with Magnetic Fields

Having described three methods we have of characterizing transition strengths, it's time to start talking about *why* these kinds of measurements are interesting in the first place. Of course, as means to an end, these kinds of measurements are necessary in order to plan efficient routes toward producing large samples of molecules in rovibrational ground states of our choosing. But since transition strengths are intimately related to the shapes of the wavefunctions describing the states we're interrogating, these measurements can additionally serve as a window into molecular structure. In particular, the degree to which transition strengths change with applied fields can tell us something about how well-approximated by traditional, selection rule-preserving molecular symmetry labels the rovibrational levels really are. Additionally, knowledge of how transition pathways are affected by the presence of external fields can be used as a tool for building new atomic and molecular clocks [13].

The following sections describe the results of our measurements of how transition strengths are affected by modest magnetic fields, as well as offer a simple model for understanding the qualitative behavior of forbidden transitions becoming allowed due to the application of small fields. While these results were obtained specifically through study of ^{88}Sr$_2$, they should be broadly applicable to many other systems. In particular, the idea that the sensitivity of transition strengths to external fields is proportional to the level spacing (implying that molecular transitions can be manipulated millions of times more strongly than atomic transitions) is very general.

5.3 Enabling "Forbidden" Transitions with Magnetic Fields

5.3.1 Intuitive Model Based on Perturbation Theory

Consider a transition from a ground state $|\gamma\rangle$ to an excited state $|\mu\rangle$. As was discussed in Sect. 5.2.2.2, the transition strength should be proportional to Rabi frequency $\Omega_{\gamma\mu}$ squared, i.e.

$$Q \propto \Omega_{\gamma\mu}^2 = \left|\frac{1}{\hbar}\langle\gamma|\hat{H}_{int}|\mu\rangle\right|^2, \qquad (5.26)$$

where $\hat{H}_{int}$ is the perturbation induced by the laser.

If the molecule is subjected to an external field capable of perturbing the molecule's energy levels, then perturbation theory tells us that a state $|\alpha\rangle$ will be modified according to

$$|\alpha(B)\rangle \approx |\alpha(0)\rangle + \sum_{\alpha \neq \nu}(B/B_{\alpha\nu})|\nu(0)\rangle, \qquad (5.27)$$

where the characteristic field $B_{\alpha\nu} = (E_\alpha - E_\nu)/\langle\alpha|\frac{\hat{H}_Z}{B}|\nu\rangle$ gives the admixing per unit field B for a pair of states $|\alpha\rangle$ and $|\nu\rangle$ which are coupled by the perturbation Hamiltonian $\hat{H}_Z$ induced by the field B. In this thesis, we're interested in applied *magnetic* fields and the associated Zeeman Hamiltonian $\hat{H}_Z = \mu_B(g_L\hat{L} + g_S\hat{S}) \cdot \vec{B}$.

For transitions from levels in the electronic ground state to those in the electronic excited state, we can use Eq. (5.27) to approximate the dependence of the transition strength on magnetic field by calculating the Rabi frequency:

$$|\Omega_{\gamma\mu}(B)|^2 \approx |\Omega_{\gamma\mu}(0)|^2 + B^2\left|\sum_{\nu\neq\mu}\frac{\Omega_{\gamma\nu}(0)}{B_{\mu\nu}}\right|^2 + B\sum_{\nu\neq\mu}\left(\frac{\Omega_{\gamma\mu}(0)\Omega_{\gamma\nu}^*(0)}{B_{\mu\nu}^*} + \frac{\Omega_{\gamma\mu}^*(0)\Omega_{\gamma\nu}(0)}{B_{\mu\nu}}\right), \qquad (5.28)$$

where I have made the assumption that the initial state $|\gamma\rangle$ is unperturbed by the magnetic field. For the case of ^{88}Sr$_2$ this assumption is a good one, since the ground state is non-magnetic (see Chap. 4).

Equation (5.28) is quite general: the only bit of molecular physics we've employed is the requirement that the ground state be insensitive to perturbations from the applied field. Let's now consider the implications for the strengths of *forbidden transitions*, i.e. those for which $\Omega_{\gamma\mu}(0) = 0$. In the case of electric dipole transitions, these would involve transitions for which $\Delta J > 1$ or $J = J' = 0$. Plugging $\Omega_{\gamma\mu}(0) = 0$ into Eq. (5.28) reduces the expression to a single term:

$$|\Omega_{\gamma\mu}(B)|^2 \approx B^2\left|\sum_{\nu\neq\mu}\frac{\Omega_{\gamma\nu}(0)}{B_{\mu\nu}}\right|^2 \qquad (5.29)$$

Equation (5.29) implies that the strength of a "forbidden transition" from $|\gamma\rangle$ to $|\nu\rangle$ will increase quadratically with the applied magnetic field, *so long as there* *exist characteristic fields $B_{\mu\nu}$ which are not too large.* This caveat is equivalent to requiring that there exist "nearby" levels $|\nu\rangle$ for which the matrix element $\langle\mu|\hat{H}_Z|\nu\rangle \neq 0$.

When will these conditions be met? Well, consider that the Zeeman Hamiltonian $\hat{H}_Z$ couples rovibrational levels with $\Delta m = 0$ and $\Delta J = 0, \pm 1$ (but not $J = J' = 0$). Then clearly transitions satisfying $\Delta J = 2$ can be described by this model, since while such transitions are E1-forbidden, nearby levels satisfying $\Delta J = 1$ with respect to the initial state can be mixed into the final state with the Zeeman Hamiltonian. A transition satisfying $\Delta J = 3$, however, would *not* be described simply by this first-order model, since the Zeeman Hamiltonian could at best only mix in levels satisfying $\Delta J = 2$ with respect to the initial state. Such a forbidden transition, with $\Delta J > 2$, would require a higher order description than is provided by Eq. (5.29). For a visual explanation of this process, see Fig. 5.2.

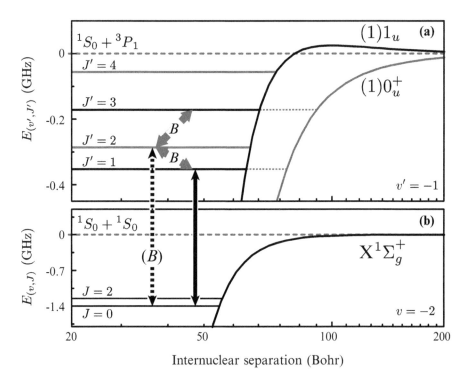

Fig. 5.2 A diagram illustrating how magnetic field mixing can cause forbidden transitions between rovibrational levels in the (**a**) excited state and (**b**) ground state to become allowed with the application of small magnetic fields. The dashed arrow represents the probe laser, while the solid arrow represents the effective transition after the final state has become admixed with $J' = 1$ character. Note that while in this simple picture only two levels are mixed into the final state, the full ab initio calculations include $J' \leq 6$, and $|\Omega'| = 0, \pm 1$. Adapted from [6]

5.3.1.1 Higher Order Terms

For transitions satisfying $\Delta J = 3$, we can go up to second order perturbation theory to get a sense of the behavior at small magnetic fields. Expanding the field-dependent wavefunction $|\mu(B)\rangle$ to second order, we have:

$$|\mu(B)\rangle = |\mu(0)\rangle + B \sum_{k \neq \mu} |k(0)\rangle \frac{\langle k(0)|\frac{\hat{H}_Z}{B}|\mu(0)\rangle}{E_\mu - E_k} \tag{5.30}$$

$$+ B^2 \left\{ \sum_{k \neq \mu} \sum_{l \neq \mu} |k(0)\rangle \frac{\langle k(0)|\frac{\hat{H}_Z}{B}|l(0)\rangle \langle l(0)|\frac{\hat{H}_Z}{B}|\mu(0)\rangle}{(E_\mu - E_k)(E_\mu - E_l)} \right. \tag{5.31}$$

$$- \sum_{k \neq \mu} |k(0)\rangle \frac{\langle \mu(0)|\frac{\hat{H}_Z}{B}|\mu(0)\rangle \langle k(0)|\frac{\hat{H}_Z}{B}|\mu(0)\rangle}{(E_\mu - E_k)^2} \tag{5.32}$$

$$\left. - \frac{1}{2}|\mu(0)\rangle \sum_{k \neq \mu} \frac{|\langle k(0)|\frac{\hat{H}_Z}{B}|\mu(0)\rangle|^2}{(E_\mu - E_k)^2} \right\}. \tag{5.33}$$

Keeping in mind that $\Delta J = 0, 1$ for E1 transitions, and that the Zeeman Hamiltonian $\hat{H}_Z$ connects states with $\Delta J = 0, 1$ as well, we can calculate the approximate transition strength $|\Omega_{\gamma\mu}^{\Delta J=3}(B)|^2$. With a bit of tedious algebra, it's straightforward to show that only a single term survives. The result is the following:

$$|\Omega_{\gamma\mu}^{\Delta J=3}(B)|^2 = B^4 \left| \sum_{k \neq \mu} \sum_{l \neq \mu} \Omega_{\gamma k}(0) \frac{\langle k(0)|\frac{\hat{H}_Z}{B}|l(0)\rangle \langle l(0)|\frac{\hat{H}_Z}{B}|\mu(0)\rangle}{(E_\mu - E_k)(E_\mu - E_l)} \right|^2. \tag{5.34}$$

Therefore for $\Delta J = 3$ transitions we would expect that at small fields, the lowest order contributions to the transition strength would be fourth-order in B.

In fact, we have been able to observe transitions satisfying $\Delta J = 3$, but unfortunately not at small enough fields to see unambiguously quartic dependence. However, ab initio calculations of the full transition strength dependence on magnetic field performed by our collaborators confirm this behavior at small fields [6]. Figure 5.3 shows theory ab initio calculations for both forbidden and "extra-forbidden" ($\Delta J = 3$) transition strengths vs magnetic field, as well as parabolic and quartic curves to guide the eye and to give a sense of when even higher order perturbation theory is necessary.

5.3.2 Results

Figure 5.4 shows several examples of measurements we've made of transition strengths for a representative sample of both forbidden and allowed transitions

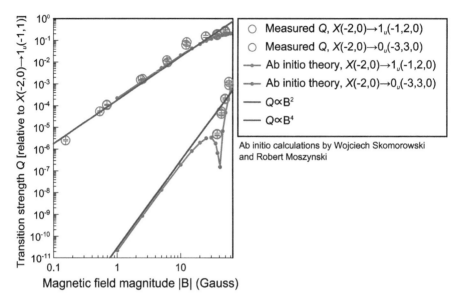

Fig. 5.3 At small magnetic fields, the strengths of "singly-forbidden" transitions (such as $X(-2,0) \rightarrow 1_u(-1,2,0)$) are expected to increase quadratically with magnetic field. "Doubly-forbidden" transitions (such as $X(-2,0) \rightarrow 0_u(-3,3,0)$) are expected to increase with the fourth power of magnetic field. Here are shown experimental data (taken from Fig. 5.4) and theoretical ab initio calculations. The solid lines represent simple quadratic and quartic fits to the data. While a quadratic fit to the strength of $X(-2,0) \rightarrow 1_u(-1,2,0)$ is fairly accurate for all measured points, the strength of $X(-2,0) \rightarrow 0_u(-3,3,0)$ is markedly non-quartic at the fields at which we make observations

across a range of applied magnetic fields. All transition strength measurements in this case were made by measuring the normalized area under a spectroscopic lineshape, as described in Sect. 5.2.1.

Because of the difficulty of determining our spectroscopy laser's beam waist, we did not measure *absolute* strengths, but rather *relative* strengths as compared to a reference E1-allowed transition which was not expected to vary much with magnetic field (see Fig. 5.4b). The "calibration measurement" shown in Fig. 5.4b was made on September 11, 2013, and all other transition strength measurements shown in Fig. 5.4 were made within approximately 2 weeks and without altering the laser optics. The different behaviors shown in Fig. 5.4 are interesting enough to merit some further discussion:

- The $\Delta J = 1$ transition depicted in Fig. 5.4b is relatively constant for all fields. For this reason, this transition was chosen as a "calibration measurement." All other transition strengths shown in Fig. 5.4 are defined relative to this allowed $\Delta J = 1$ transition.
- In the most extreme cases, we demonstrate control of transition strengths over a range of more than five orders of magnitude with field magnitudes of only a few tens of Gauss, with some strengths becoming comparable in magnitude to

5.3 Enabling "Forbidden" Transitions with Magnetic Fields

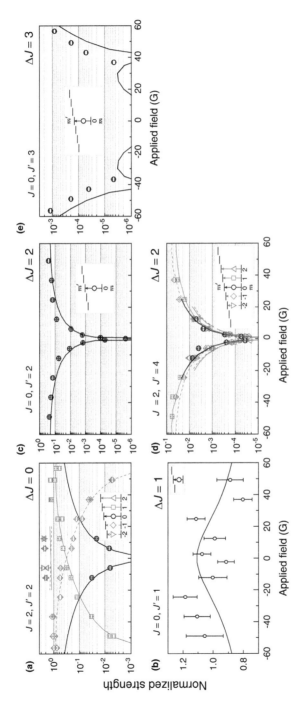

Fig. 5.4 With the application of small magnetic fields, the strengths of forbidden transitions can be modified dramatically, in some cases becoming comparable to those of E1-allowed transitions. The transitions shown are from $X(-2, J, m)$ to (**a**) $1_u(-1, 2, m')$, (**b**) $1_u(-1, 1, 0)$, (**c**) $1_u(-1, 2, 0)$, (**d**) $1_u(-1, 4, m')$, and (**e**) $0_u(-3, 3, 0)$. Adapted from [6]

the allowed $\Delta J = 1$ transition. This is dramatically larger than the control which can be achieved by analogous atomic transitions, and is due to the very dense level spacing of weakly bound molecules as compared to atoms.

- In Fig. 5.4c, d, describing "forbidden" transitions for which $\Delta J = 2$, the dependence of the transition strength upon magnetic field is approximately quadratic, as we would expect from our first-order perturbation theory arguments in the previous section. The $m = 0$ component of Fig. 5.4a also increases quadratically with field. This is because the $(J = 2, m = 0) \rightarrow (J' = 2, m' = 0)$ transition is "accidentally" forbidden due to a vanishing Clebsch-Gordon coefficient.
- The dramatic $m' \pm 1$ asymmetry seen in Fig. 5.4a is a result of interference effects induced by admixing. This interference and linear magnetic field dependence can be seen as a consequence of the third term of Eq. (5.28), and is present to a smaller degree in Fig. 5.4d as well.

We believe that similar physics is at play in governing the behavior of the linewidths of transitions to subradiant states in the presence of magnetic fields, since such transitions are observed to narrow linearly with field in some cases and broaden quadratically in others. For details, see Chap. 6.

References

1. Einstein, A.: Strahlungs-emission und absorption nach der quantentheorie. Dtsch. Phys. Ges. **18**, 318–323 (1916)
2. Foot, C.: Atomic Physics. Oxford Master Series in Atomic, Optical, and Laser Physics. Oxford University Press, Oxford (2005)
3. Hilborn, R.: Einstein coefficients, cross sections, f values, dipole moments, and all that (2002). arXiv preprint physics/0202029
4. Ido, T., Katori, H.: Recoil-free spectroscopy of neutral Sr atoms in the Lamb-Dicke regime. Phys. Rev. Lett. **91**(5), 053001 (2003)
5. Martinez de Escobar, Y., Mickelson, P., Pellegrini, P., Nagel, S., Traverso, A., Yan, M., Côté, R., Killian, T.: Two-photon photoassociative spectroscopy of ultracold ^{88}Sr. Phys. Rev. A **78**, 062708 (2008)
6. McGuyer, B., McDonald, M., Iwata, G., Skomorowski, W., Moszynski, R., Zelevinsky, T.: Control of optical transitions with magnetic fields in weakly bound molecules. Phys. Rev. Lett. **115**(5), 053001 (2015)
7. McGuyer, B., McDonald, M., Iwata, G., Tarallo, M., Grier, A., Apfelbeck, F., Zelevinsky, T.: High-precision spectroscopy of ultracold molecules in an optical lattice. New J. Phys. **17**(5), 055004 (2015)
8. McGuyer, B., McDonald, M., Iwata, G., Tarallo, M., Skomorowski, W., Moszynski, R., Zelevinsky, T.: Precise study of asymptotic physics with subradiant ultracold molecules. Nat. Phys. **11**(1), 32–36 (2015)
9. Metcalf, H., van der Straten, P.: Laser Cooling and Trapping. Graduate Texts in Contemporary Physics. Springer, New York (1999)
10. Osborn, C.: The physics of ultracold Sr$_2$ molecules: optical production and precision measurement. PhD Thesis (2014)
11. Reinaudi, G., Osborn, C., McDonald, M., Kotochigova, S., Zelevinsky, T.: Optical production of stable ultracold ^{88}Sr$_2$ molecules. Phys. Rev. Lett. **109**, 115303 (2012)

12. Skomorowski, W., Pawłowski, F., Koch, C., Moszynski, R.: Rovibrational dynamics of the strontium molecule in the $A^1\Sigma_u^+$, $c^3\Pi_u$, and $a^3\Sigma_u^+$ manifold from state-of-the-art *ab initio* calculations. J. Chem. Phys. **136**(19), 194306 (2012)
13. Taichenachev, A., Yudin, V., Oates, C., Hoyt, C., Barber, Z., Hollberg, L.: Magnetic field-induced spectroscopy of forbidden optical transitions with application to lattice-based optical atomic clocks. Phys. Rev. Lett. **96**(8), 083001 (2006)

Chapter 6
Subradiant Spectroscopy

6.1 Introduction: Subradiance vs Superradiance

When a single atom undergoes a transition to an unstable excited state, it will eventually decay. The exact time at which that atom will decay is impossible to predict, but we can quantify the *probability* that after a certain time t the atom will have decayed to its ground state with the following formula:

$$P(\text{decay}) = 1 - e^{-\Gamma t} \tag{6.1}$$

The quantity Γ is the *radiative decay rate*, and its inverse $\tau = \dfrac{1}{\Gamma}$ is the state's lifetime, i.e. the amount of time it takes for the probability of decay to reach $1-1/e$ ($\sim$63%).

The molecules described in this thesis can be thought of as two atoms "glued together," one in the stable ground state and one in the unstable excited state. As a first guess at the properties of such molecules, we might try to reason by analogy with atoms. We would expect such an analogy to be imperfect, but perhaps to get better and better as the molecular bond length increases. But in many cases, such as was described in Chap. 4 concerning the behavior of quadratic Zeeman shifts, the analogy breaks down completely, and we're forced to reckon with molecules in their full quantum mechanical glory. The decay rates of unstable molecular states are no exception.

"Subradiance" and "Superradiance" are terms describing inherently quantum mechanical effects, arising from the collective interactions of several particles. The terms first appeared in the scientific literature in a 1953 paper by Robert Dicke [5], and denote the suppression or amplification, respectively, of spontaneous radiation. When particles are closer together than the wavelength of their emitted radiation, interference between the radiation of nearby particles becomes important.

Depending on the initial quantum state of the gas, the interference can be either constructive, leading to an amplification of the decay rate, or destructive, causing the decay rate to approximate zero.

Historically, superradiance has proven far easier to study than subradiance, since for a system exhibiting both effects, superradiance will dominate observations. In molecules, there are few examples of precise studies of subradiance. Subradiant states have been observed in ytterbium [15], though careful measurements of the decay rates at the time were impossible. Suppression and amplification of spontaneous emission from a pair of nearby ions has also been demonstrated [4], though the magnitude of the effect was small owing to the relatively large particle separation. Recently, it has been demonstrated that it is possible to observe both many-body subradiance and superradiance by monitoring the time evolution of spontaneous radiation from a specially prepared atomic cloud [6].

In this chapter I'll describe work published in 2015 [9] describing our observations of highly subradiant states in $^{88}Sr_2$, which were enabled by several factors relatively unique to our experiment. First, our ability to produce and probe ultracold molecules in an optical lattice allows for high resolution, Doppler-free spectroscopy and quantum state control. Fine spectroscopic resolution is essential for characterizing narrow linewidths, while precise quantum state control is necessary for engineering conditions such that electric dipole (E1) transitions are forbidden, leaving only higher order magnetic dipole (M1) and electric quadrupole (E2) transitions for accessing subradiant states. Second, theoretical understanding of the $^{88}Sr_2$ molecule has advanced enough to allow for ab initio predictions of many molecular properties [12–14] such as binding energies, state lifetimes, and transition strengths from the ground state. Calculations by Robert Moszynski and Wojciech Skomorowski at the University of Warsaw were critical to the discovery and eventual understanding of these states.

6.2 Characterizing Transition Strengths

Because M1 and E2 interactions are so much weaker than E1 transitions, the very first question we should ask is *how much weaker*. Answering this is critically important, because it influences whether or not we have any hope of finding subradiant states in $^{88}Sr_2$, and informs the strategy we should choose when searching for them. For E1, M1, and E2 transitions, we define dimensionless oscillator strengths f_{12} in the following way:

$$f_{12}^{E1} = \frac{2m_e \omega_{12}}{g_1 \hbar} \sum_{M'} |\vec{\epsilon} \cdot \langle X0_g^+, v, J, M | \hat{\mathbf{r}} | 1_g, v', J', M' \rangle|^2 \tag{6.2a}$$

$$f_{12}^{M1} = \frac{2m_e c}{g_1 \hbar e^2} \sum_{M'} |(\vec{e}_k \times \vec{\epsilon}) \cdot \langle X0_g^+, v, J, M | \hat{\boldsymbol{\mu}} | 1_g, v', J', M' \rangle|^2 \tag{6.2b}$$

6.2 Characterizing Transition Strengths

$$f_{12}^{E2} = \frac{m_e \omega_{12}^2}{2g_1 \hbar c} \sum_{M'} |\vec{\epsilon} \cdot \langle X0_g^+, v, J, M|\hat{Q}|1_g, v', J', M'\rangle \cdot \vec{e}_k|^2 \tag{6.2c}$$

where $\hat{r}$, $\hat{\mu}$, and $\hat{Q}$ are the electric dipole, magnetic dipole, and electric quadrupole operators, respectively (defined as in [16]), $\vec{\epsilon}$ is the polarization vector, $\vec{e}_k$ is the dimensionless unit wave vector of the light, g_1 is the degeneracy of the initial state, ω_{12} is the angular frequency of the transition (nearly equal for all weakly-bound states), and all other constants should be self-evident.

Note that the E1, M1, and E2 operators are functions of the local field intensity, which is notoriously difficult to measure precisely. This difficulty in fact prevents us from making accurate determinations of the *absolute* transition strengths of these states. However, it is much easier to guarantee that even if the total field intensity is unknown, it is the same for a series of measurements of different states. Therefore, rather than measure *absolute uncertainties*, we measure *relative uncertainties* (as discussed in Chap. 5). The following sections give the results of our measurements of transition strengths via two complementary methods.

6.2.1 Aside: Isolating E1, M1, and E2 Transitions from One Another

For a given transition $\psi_i \to \psi_f$, it may be the case that there are several allowed pathways (e.g., both M1 and E2). This situation would result in the measured strength of the transition being the superposition of two contributions, which would make comparison with theory more difficult and less precise. To avoid this scenario, we make use of selection rules to guarantee that one and only one pathway from among the choices of E1, M1, or E2 is allowed, so that the transition strength we measure is due purely to a single channel. For the data presented in this chapter, we study the following cases, which each guarantee that only the listed transition pathway is allowed:

- **E1 transitions**
  Vertical magnetic field (i.e., quantization axis), vertical laser polarization, studying transitions from (**gerade,** $J = 0, m = 0$)→(**ungerade,** $J' = 1, m' = 0$)
- **M1 transitions**
  Vertical magnetic field (i.e., quantization axis), horizontal laser polarization, studying transitions from (**gerade,** $J = 0, m = 0$)→(**gerade,** $J' = 1, m' = 0$)
- **E2 transitions**
  Vertical magnetic field (i.e., quantization axis), horizontal laser polarization, studying transitions from (**gerade,** $J = 0, m = 0$)→(**gerade,** $J' = 2, m' = \pm 1$)

For the case of E2 transitions, which allow transitions between both $\Delta m = +1$ and $\Delta m = -1$, we measure the total strength to both spectroscopic peaks and then take the average of the two.

6.2.2 Normalized Area Under a Lorentzian

This method of measuring the relative transition strength involves recording high-resolution spectroscopic traces of transitions to subradiant states, and then normalizing the area under the trace to the product of (probe time)×(probe power), as was discussed in detail in Chap. 5. Because of the extremely narrow linewidths of these transitions, we were concerned that power-broadening could have an especially adverse effect on these measurements as compared to transitions to superradiant states. We therefore recorded traces at a series of probe powers in order to search for evidence of power-broadening or obvious outliers, taking a weighted average of only those traces exhibiting no obvious relationship between the probe power used and the linewidth of the state.

Figure 6.1 shows representative data sets describing transition strength measurements for E1, M1, and E2 transitions from the initial state $X(-1, 0)$. In order to define a standard transition strength reference, measurements of transitions to the $1_u(-1, 1)$ state were made first. This is shown in part (a). Immediately afterward, the transitions to subradiant states were studied at various probe powers. In all cases, a Lorentzian function is fit to the *natural log* of the data, as described in Chap. 5. So as not to bias the fit with noisy, low signal data points, error bars are added to each point according to $\Delta[\ln(signal)] = \Delta(signal)/(signal)$, where $\Delta(signal)$ was estimated as the shot-to-shot noise on the signal.

6.2.3 Rabi Oscillations

The exceptionally long lifetimes of subradiant states in ^{88}Sr$_2$ enable Rabi oscillations to become visible in the excited state population after interaction with a probe laser of finite duration. Since the rate at which the population oscillates between ground and excited state is proportional to both the square root of the intensity of the probe and the transition dipole moment of the transition, a measurement of the Rabi-flopping rate can be used to determine transition strength, as was described in detail in Chap. 5.

Figure 6.2 shows measurements of Rabi oscillations between $X(v, 0)$ and $1_g(v', 1, 0)$ states, as well as the dependence of Rabi frequency upon probe power. Only the $J' = 1$ 1_g states were probed in this way because only these states appear to be narrow enough to allow for coherent manipulations at long (several hundred μs) timescales. In part (a), the Rabi frequencies Ω for transitions to the four $1_g(v', 1, 0)$ states from $X(-1, 0)$ or $X(-2, 0)$ are shown at various probe powers, and all fall along curves given by $\Omega(P) = A \cdot P^{\frac{1}{2}}$. On a log-log plot, this means that all curves will have the same slope, and a larger value of A implies a larger vertical offset of the curve. These Rabi frequencies plotted in this way are determined by fitting the data shown in part (b) to the following equation:

6.2 Characterizing Transition Strengths

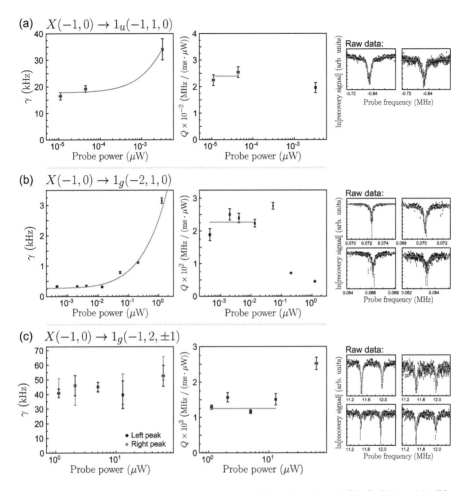

Fig. 6.1 Here are shown three example data sets, used for calculating (**a**) E1, (**b**) M1, and (**c**) E2 transition strengths from the starting state $X(-1, 0)$. For each data set, a spectrum was taken at several probe powers. Then the quantity (area)/(power×time) was calculated for each trace (see text). To arrive at the final "transition strength" Q, values at low probe power (i.e., those displaying no obvious power broadening) were combined in a weighted average. Since we are interested in the area of a Lorentzian fit the log() of the signal, error bars in the "Raw data" images were defined as $\Delta[\ln(signal)] = \Delta(signal)/(signal)$ before fitting. $\Delta(signal)$ was estimated as the shot-to-shot noise on the signal

$$N(t) = y_0 e^{-\frac{\Gamma_1}{2}t} + A e^{-\frac{\Gamma_2}{2}t} \cos\left[\frac{1}{2}\sqrt{\Omega^2 - \frac{\Gamma_2^2}{4}} \cdot (t - t_0)\right], \quad (6.3)$$

where for all data except that which describes transitions to $1_g(-1, 1, 0)$ we set $\Gamma_1 = 0$.

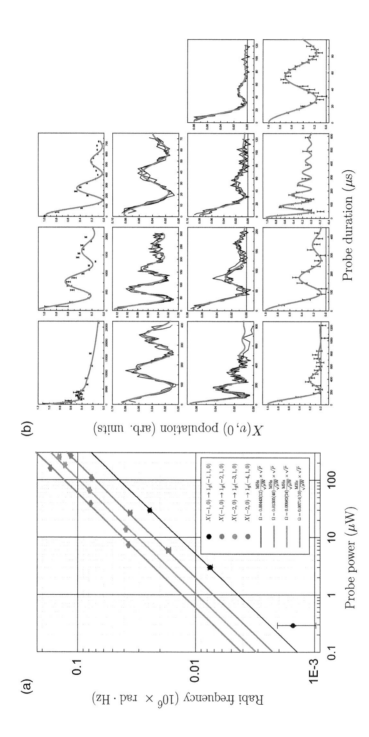

Fig. 6.2 (**a**) A plot of the Rabi flopping frequency versus probe power, showing a square-root dependence on power for each transition. (**b**) Raw data showing Rabi flopping between initial and final states. Traces are ordered from low power (left) to high power (right). From top to bottom, the transitions shown are (1) $X(-1, 0) \to 1_g(-1, 1, 0)$, (2) $X(-1, 0) \to 1_g(-2, 1, 0)$, (3) $X(-2, 0) \to 1_g(-3, 1, 0)$, and (4) $X(-2, 0) \to 1_g(-4, 1, 0)$. The red curve is a fit described in the text

6.2 Characterizing Transition Strengths

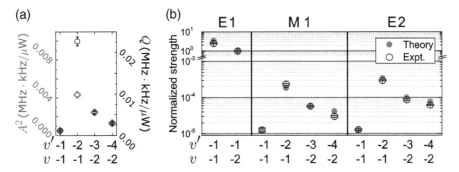

Fig. 6.3 (a) A comparison of Rabi flopping-derived (red points/left axis) and spectroscopy-derived (black points/right axis) measurements of the transition strengths of transitions from $X(v,0)$ to $1_g(v,1,0)$. Note that while the units for both measurements are the same, the relative scaling has been adjusted to allow for easier relative comparison. (b) Published relative transition strength values (black points) derived from spectroscopy measurements, as well as comparison with ab initio theory calculations (red points) (adapted from [9])

Surprisingly, the state with the longest lifetime, $1_g(-1,1,0)$, appears to exhibit damping of its Rabi oscillations at a rate much faster than its natural decay rate of ~ 30 Hz. This is due to an experimental quirk, in which the same laser driving the transition from $X(-1,0)$ to $1_g(-1,1,0)$ is simultaneously energetic enough to dissociate the newly created $1_g(-1,1,0)$ molecules above the $^3P_1+^3P_1$ threshold (see, e.g., Fig. 3.6). The short coherence times for the remaining data sets are less well-understood.

The dependence of the Rabi frequency on probe power can be used to extract information about the transition strengths. As was described in detail in Chap. 5, we would expect that the transition strength Q should be related to the Rabi frequency ω_R via

$$Q = \frac{1}{4}\left(\frac{\omega_R^2}{P}\right), \quad (6.4)$$

where P is the laser power used to produce a Rabi frequency ω_R. Alternatively, since the dependence of Rabi frequency upon probe power can be written $\omega_R = A \cdot P^{1/2}$, we can write:

$$Q = \frac{A^2}{4}. \quad (6.5)$$

Figure 6.3a compares transition strength measurements made spectroscopically to those made with Rabi flopping. Because of a stubborn disagreement in the overall scaling between the two methods, the data plotted in Fig. 6.3a has been scaled so as to allow relative comparison between the two data sets (see the differently colored axes at left and right). One possible reason for this disagreement in the overall scaling would be a slight drift in the alignment of the probe laser with the atom cloud. This is a hard problem to completely eliminate. All spectroscopy-derived transition strength measurements presented in this chapter were made over the course of 2 days (from February 5, 2014 to February 6, 2014), while the Rabi-flopping measurements were made over the course of 5 days, but nearly 2 weeks later. While each set of measurements should be self-consistent, a small bump of a mirror during the intervening time could potentially cause disagreement between the two sets. The relative disagreement of the $1_g(-2, 1, 0)$ state with its companions is harder to explain away, and as of the writing of this thesis still remains somewhat mysterious.

Figure 6.3b shows plots of *relative* transition strengths, normalized to the value of the transition strength of an E1 transition to the $1_u(-1, 1, 0)$ state. For the data presented here (and published in [9]), our data is derived purely from spectroscopic measurements because of some lingering doubts about our ability to accurately interpret our Rabi frequency data. However, because of the jarring disagreement of the Rabi frequency data for the $1_g(-2, 1, 0)$ state, we have increased the error bar for that datum so that the two techniques agree to within 2σ.

6.3 Characterizing Linewidths (I): Sources of *Artificial* Broadening

After finding subradiant states and characterizing the strengths of transitions to them from the electronic ground state, we can begin to ask more detailed questions about what makes them special. Perhaps the most obviously interesting quantity is their very long lifetimes and narrow transition linewidths, since the hallmark of subradiance (as opposed to superradiance) in singly-excited homonuclear diatomic molecules is a transition linewidth narrower than twice the linewidth of the atomic transition [5, 15].

We use two methods for measuring linewidths which complement one another: spectroscopy, suitable for broad transitions to shorter-lived states, and "in the dark" lifetime measurements, better suited for narrow transitions to long-lived states. Experimental imperfections can plague each of these methods in different ways, and need to be properly accounted for in order to reveal the true linewidth of a transition. The following sections describe a few of the experimental issues we've uncovered, and the tricks we've used to minimize them in order to ensure that our measurements reflect the true natural linewidths of the transitions under investigation.

6.3.1 *(I) Spectroscopy*

"Spectroscopy" refers to sweeping a laser across a molecular transition and recording its lineshape, i.e. the excitation probability as a function of laser frequency.

6.3 Characterizing Linewidths (I): Sources of *Artificial* Broadening

A lineshape recorded in this way will reflect the transition's "natural linewidth" only if all other artificial sources of broadening are significantly smaller than the transition's natural linewidth $\gamma = \frac{1}{2\pi\tau}$, where τ is the $\frac{1}{e}$ exponential decay lifetime of the excited state. Here are some of the most important issues which can contribute broadening larger than the natural linewidth if not carefully controlled.

6.3.1.1 Blurring Due to a "Messy Probe"

A measurement of any experimental quantity can only be as good as the tool used to measure it. For spectroscopy our tool is a narrow-linewidth extended cavity diode laser (ECDL), stabilized to a high-finesse cavity. Several factors can combine to artificially broaden our probe laser.

Finite Interrogation Pulse Time ("Fourier Broadening") No finite laser pulse can have exactly zero linewidth. This can be thought of as a result of the uncertainty principle: a finite measurement time implies a finite uncertainty governing the photon energy (or frequency). Alternatively, it can also be derived simply as a result of Fourier decomposition. For example, for a laser pulse which is discretely switched on and off via a fast AOM (as we do in our experiment), the electric field of the laser at the location of the molecules being probed can be modeled as the product of a cosine wave times a box function:

$$E(t) = \begin{cases} E_0 \cdot \text{Re}[e^{(2\pi i f_0 t)}] & t \in (0, T) \\ 0 & \text{otherwise} \end{cases} \quad (6.6)$$

The spectral composition of this function is given by its Fourier transform, which in this case is a sinc function centered at $f = f_0$. Since the transition probability is proportional to laser *intensity* rather than electric field, the lineshape for a Fourier-limited square pulse will be proportional to a sinc^2 function.

Figure 6.4 gives a visual illustration of this effect, as well as experimental data proving its very real existence. The data shown in Fig. 6.4 depicts spectroscopy of a transition from $X(-1, 0) \rightarrow 1_g(-1, 1, 0)$. For this measurement the pulse time was chosen to be very small (50 μs) in order to reveal Fourier broadening as the dominant line-broadening mechanism. Note that the Fourier-broadened linewidth $\gamma_{FB} \approx \frac{1}{T}$, and therefore revealing the natural linewidth of 30 Hz would require a pulse duration of $T \gtrsim \frac{1}{30\,\text{Hz}} \approx 33$ ms. However, other effects in our experiment, namely the natural linewidth of our spectroscopy laser, limit us to useful coherence times of 5–10 ms.

Finite Laser Linewidth While Fourier broadening can be minimized simply by increasing the probe time, broadening due to the inherent linewidth of the laser can be decreased only by building a better laser. Characterizing the linewidth of a narrow laser is difficult [2]. In our experiment, however, we have access to very narrow molecular transitions, which allow us to infer the linewidth of our probe

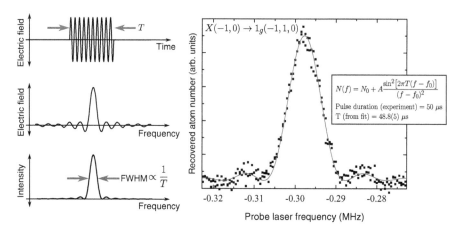

Fig. 6.4 Fourier broadening of a gated laser pulse. On the left is shown a qualitative illustration of the mathematical underpinning of Fourier broadening, while on the right is shown real data depicting spectroscopy of the $X(-1, 0) \rightarrow 1_g(-1, 1, 0)$ transition

laser operationally as the linewidth of a transition to a long-lived molecular state when all other sources of broadening have been minimized.

The black points in Fig. 6.5a depict spectroscopy of the $X(-1, 0) \rightarrow 1_g(-1, 1, 0)$ transition. The linewidth of a (properly scaled) Lorentzian fit to these points is 150(20) Hz, substantially larger than the natural linewidth of 28.5(2) Hz (determined by "in the dark" lifetime measurements described later). Several sources of broadening must be accounted for before extracting the laser linewidth.

In Chap. 7 we will learn that FWHM $\approx 0.3 \times$ (Total light shift). For the data shown, recorded at 209 mW lattice power and a measured lattice light shift of $\sim$1 Hz/mW, we would expect 209 mW $\times$ 1 Hz/mW $\times$ 0.3 $\approx$ 63 Hz of broadening due to light shifts. The 20 ms probe time used for this trace would lead to $\sim \frac{1}{20\,\text{ms}} \approx$ 50 Hz of Fourier broadening. The $\sim$20 ms collisional lifetime of ground state molecules would contribute another $\sim$50 Hz to the linewidth. Finally, the 430 mG applied magnetic field used to set the quantization axis would produce $\sim$60 Hz of natural broadening due to mixing of nearby shorter-lived states (see Fig. 6.11).

If we ignore the detailed lineshapes for each of these effects, and instead model each as a Gaussian (bell curve) with a linewidth γ_{source}, we can relate the experimentally observed linewidth γ_{exp} to the laser linewidth γ_{probe} with the following equation (which makes use of the fact that the convolution of two Gaussians with FWHM's of γ_1 and γ_2 produces another Gaussian with FWHM $\gamma_{\text{total}} = \sqrt{\gamma_1^2 + \gamma_2^2}$):

$$\gamma_{\text{exp}} = \sqrt{\gamma_{\text{probe}}^2 + \gamma_{\text{natural}}^2 + \gamma_{\text{lattice}}^2 + \gamma_{\text{Fourier}}^2 + \gamma_{\text{magnetic}}^2 + \gamma_{\text{collisional}}^2} \quad (6.7)$$

6.3 Characterizing Linewidths (I): Sources of *Artificial* Broadening

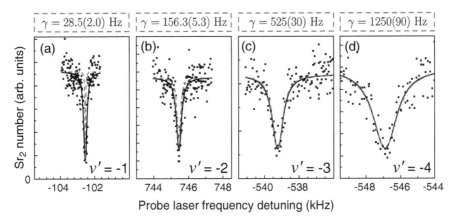

Fig. 6.5 Spectroscopic data depicting measured lineshapes for transitions (**a**) $X(-1,0) \rightarrow 1_g(-1,1,0)$; (**b**) $X(-1,0) \rightarrow 1_g(-2,1,0)$; (**c**) $X(-2,0) \rightarrow 1_g(-3,1,0)$; and (**d**) $X(-2,0) \rightarrow 1_g(-4,1,0)$. Black points represent experimental data while solid black curves represent Lorentzian fits to the data (log-scaled to account for linear probe absorption). Dashed curves represent Lorentzians with the "true" linewidth (shown in the red dashed boxes above the data), as determined by "in the dark" lifetime measurements described in Sect. 6.3.2. Adapted from [9]

$$150\,\text{Hz} = \sqrt{\gamma_{\text{probe}}^2 + (28.5\,\text{Hz})^2 + (63\,\text{Hz})^2 + (50\,\text{Hz})^2 + (60\,\text{Hz})^2 + (50\,\text{Hz})^2} \tag{6.8}$$

$$\rightarrow \gamma_{\text{probe}} \approx 95\,\text{Hz} \tag{6.9}$$

This admittedly rough calculation gives an idea of the linewidth of our probe laser, and allows us to estimate when spectroscopy is a valid tool to use for determining linewidths. Figure 6.5 shows spectroscopic lineshapes measured for transitions to 1_g states whose natural linewidths range from $\sim$30 Hz$\rightarrow$ 1.25 kHz. Clearly spectroscopy (black points and solid line) gives artificially broad results for the narrowest two transitions, but agrees nearly perfectly with results derived from lifetime measurements (dashed red curve) for the two broadest transitions. (For lifetime measurement details, see Fig. 6.7.)

6.3.1.2 Blurring Due to "Messy Molecules"

In addition to line blurring due to a messy probe, blurring due to "messy transitions," i.e. effects which cause the transition frequencies of different molecules to become shifted with respect to one another, is important to minimize. It's critical that all molecules in the trap are subject to the same perturbing environment, so that the same shift is common to all molecules. If this condition is satisfied, then transitions will be narrow, and the accuracy with which these shifts can be evaluated and subtracted is greatly improved.

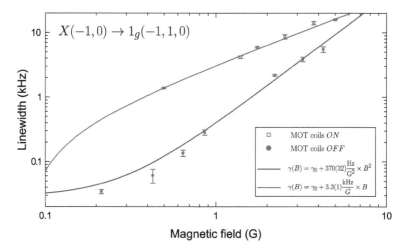

Fig. 6.6 This shows a comparison of the linewidth of the $X(-1,0) \rightarrow 1_g(-1,1,0)$ transition with the MOT coils left *ON* during probing (which introduces a magnetic field gradient across the cloud) vs with the MOT coils turned *OFF*. The "MOT coils ON" measurements were made spectroscopically, whereas the "MOT coils OFF" measurements were made either by determining the lifetime of $1_g(-1,1,0)$ molecules "in the dark" (for small widths) or spectroscopically (for large widths)

Zeeman Shift Blurring Due to Magnetic Field Gradients It is an unfortunate coincidence that the narrowest transitions to subradiant states also posses the largest quadratic Zeeman shifts (see Fig. 4.1). For a transition which shifts according to $f = f_0 + qB^2$, a blurring ΔB of the magnetic field can produce a blurring Δf of approximately:

$$\Delta f \approx \frac{df}{dB} \Delta B = 2|q|B\Delta B \qquad (6.10)$$

From the above equation, it is clear that to minimize blurring due to field gradients, one must either minimize the gradient ΔB, minimize the magnetic field B, or choose a state with a small quadratic Zeeman shift q. Quantization-axis considerations (e.g., the magnitude of the lattice tensor light shift compared to the linear Zeeman shift) place a lower limit on B, and we have no control over the quadratic Zeeman shift coefficient q. Therefore we'd like to work as hard as possible to minimize ΔB.

In order to minimize magnetic field gradients, we switched the MOT field coils off during measurements. Figure 6.6 shows measurements of the $X(-1,0) \rightarrow 1_g(-1,1,0)$ transition linewidth at various magnetic fields, with the MOT coils kept ON vs OFF. The linear broadening of $\sim$3.3 kHz/G allows us to estimate the total "magnetic field blur" experienced by the molecules being probed:

$$\Delta B \approx \frac{1}{2|q|} \times \frac{\Delta f}{B} \qquad (6.11)$$

6.3 Characterizing Linewidths (I): Sources of *Artificial* Broadening

$$= \frac{1}{2 \times 121 \text{ kHz/G}^2} \times 3.3 \text{ kHz/G} \quad (6.12)$$

$$\approx 14 \text{ mG} \quad (6.13)$$

This result is of the same order as that which we would expect from multiplying the known field gradient of our MOT coils of $\sim$11 G/cm [11] by the approximate size of our probed molecule sample of $\sim$20 μm. The small discrepancy in the two results can be attributed to not being super-careful with our definitions of "Δf" and "ΔB," as well as the magnetic field gradient not being exactly linear in only one direction.

From the figure, it's clear that any broadening due to B-field gradients is significantly smaller when the MOT coils are turned off. We worked hard to minimize any other sources of B-field blurring (both spatial and temporal) by making a careful survey of magnetic fields produced by electronics and permanent magnets near the vacuum chamber, repositioning those which were the worst emitters so that they were as far from the science chamber as possible.

Note that quantifying the magnitude of spatial and temporal field gradients near the science chamber was not a trivial task. Doing so required a probe with high relative precision, capable of making measurements of small changes in the magnetic field at fairly high fields, since we are interested in determining field noise in realistic experimental conditions (e.g., when the DC magnetic loading coils are turned on). We used a Bartington Mag-03MSESB1000 three-axis magnetometer to search for DC field gradients. For AC gradients, we build a magnetic flux probe consisting of a single loop of coax cable with the shielding split at one end.

Differential ("Non-magic") Lattice Light Shift Broadening If the initial and final molecular states see different trap depths, and if the initial molecular cloud is at non-zero temperature, then transitions will be blurred due to inhomogeneous light shifts seen by molecules at different positions and initial energies. Empirically, we've observed that for our experiment, FWHM $\approx 0.3 \times$ (total light shift). See Chap. 7 for details.

All measurements of narrow linewidths were made in lattices set to be as nearly magic as possible. Our experimental resolution enables us to determine the magnitude of a lattice light shift to a precision of $\sim$1 Hz/mW, limited mainly by the spectroscopic resolution of our probe lasers and the limited lattice power available for measurements. This value would imply for a typical lattice power of 200 mW a lattice light shift-induced blurring of $\sim$60 Hz. Future improvements to this 1 Hz/mW precision can come from better stabilization of the high finesse cavity resonance (e.g., better thermal stability), a narrower intrinsic probe laser linewidth, better stabilization of the lattice power, or an increase of the total lattice power available for trapping.

6.3.2 (II) "In the Dark" Lifetime Measurements

The most significant source of blurring described in the previous section consisted of a broad ($\sim$100 Hz) intrinsic probe laser linewidth. This sets a lower limit on the linewidth measurable by spectroscopy, and rules out studying the $\sim$30 Hz linewidth of the narrow $X(-1, 0) \rightarrow 1_g(-1, 1, 0)$ transition or the $\sim$150 Hz linewidth of the $X(-1, 0) \rightarrow 1_g(-2, 1, 0)$ transition with this technique. Instead, to measure these very narrow transitions we have developed a technique to measure state lifetimes, illustrated in Fig. 6.7. The following sections describe the precautions which must be taken in order to ensure that lifetime measurements record the true lifetimes of the molecular states under study.

6.3.2.1 Spontaneous Decay to "Visible" States (i.e., Atoms)

Our pumping and imaging scheme relies on the idea that molecules decaying from the states under investigation will decay to states invisible to imaging and inaccessible to pumping. For the $1_g(v = -2, -3, -4; J = 1)$ states this is satisfied easily, since the primary decay mechanism is *predissociation*, whereby the molecules tunnel to an unbound atomic state with large kinetic energy which quickly escapes the trap, as discussed in Sect. 6.4.2. For the $1_g(-1, 1)$ state, however, a significant fraction of molecules decay to slow-moving free atoms, which subsequently can be imaged, artificially inflating the recovery signal. This process is represented by the open squares in Fig. 6.7. Since there is no guarantee that the spontaneously decaying atoms will produce a signal with a decay constant exactly equal to that of the properly recovered atoms (and in fact, the functional form of the number of decaying atoms versus time should not even be exactly exponential), this spontaneous decay signal must be subtracted from the "true" signal before fitting with an exponential decay.

6.3.2.2 Imperfect π-Pulse Generation

In order to measure the lifetime of a molecular state, significant population must first be pumped into that state, held for a variable length of time, and then reliably retrieved for imaging. We achieve high-efficiency state transfer with Rabi π-pulses, as shown in Fig. 6.7a. However, we must be careful: an imperfect choice for the duration for a π-pulse can dramatically affect the measured value for the state lifetime. If the duration of the "π-pulse" differs significantly from the "true value," the "recovery pulse" will sample a superposition of initial and final states rather than a pure state, and as a result the number of recovered molecules will vary as a function of "wait time" in a way which is unrelated to the excited state's natural lifetime.

6.3 Characterizing Linewidths (I): Sources of *Artificial* Broadening

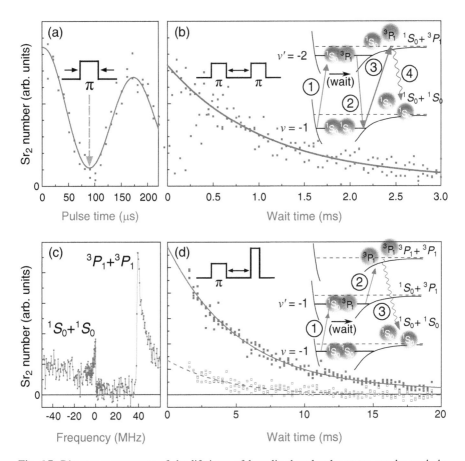

Fig. 6.7 Direct measurements of the lifetimes of long-lived molecular states can be made by coherently pumping ground state molecules into the 1_g state under consideration, waiting for a variable amount of time, and then measuring the amount of population remaining in the 1_g state. (**a**) Coherent Rabi oscillations are observed by measuring the population of $X(-1,0)$ after being subjected to a probe laser (resonant with the $X(-1,0) \rightarrow 1_g(-1,1,0)$ transition) of variable duration. The duration of the pumping pulse is chosen as the smallest value which minimizes the number of $X(-1,0)$ molecules observed. (**b**) Data depicting the number of molecules observed in the $1_g(-2,1,0)$ state as a function of wait time, along with an illustration outlining the experimental sequence in more detail. (**c**) Molecule population in the $1_g(-1,1,0)$ state can be measured by photodissociating via the $^1S_0+^1S_0$ (left) or $^3P_1+^3P_1$ (right) thresholds. (**d**) Data depicting the number of molecules observed in the $1_g(-1,1,0)$ state as a function of wait time, with an illustration describing the modified experimental sequence. Note that because $1_g(-1,1,0)$ molecules can spontaneously decay to free atoms, this signal (which can be measured simply by omitting steps 2 and 3 in the illustration) must be subtracted from the total recovery signal in order to determine the true lifetime. Adapted from [9]

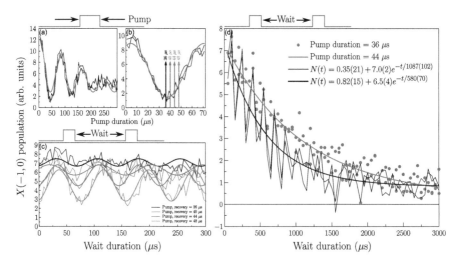

Fig. 6.8 (a) A sample of molecules is initially produced in the $X(-1, 0)$ state, and then subjected to a pump laser resonant with the $1_g(-2, 1, 0)$ state of variable duration. Afterward, the number of $X(-1, 0)$ molecules remaining is imaged. This number exhibits Rabi oscillations as a function of pump time. (b) The same experimental sequence as that which is depicted in the previous panel, but with finer time resolution. Arrows mark the pump durations which will be used later. (c) A full "pump, wait, recover" sequence is shown for four different choices of "pump/recovery" duration. When the pump duration differs significantly from the optical π-pulse value of $\sim$36 μs, large oscillations in the recovered molecule number develop due to sampling of a time-evolving superposition of initial and final states rather than a single pure state. (d) Measurements of the $1_g(-2, 1, 0)$ state lifetime with the "correct" π-pulse duration of 36 μs vs the "incorrect" π-pulse duration of 44 μs. An exponential fit to each data set yields decay constants which differ by nearly a factor of two

Figure 6.8 shows this effect at play in a measurement of the lifetime of the $1_g(-2, 1, 0)$ state. Since state lifetime is determined by fitting a plot of the molecule population vs time with an exponential decay function, the fitted value for the decay constant τ can be strongly affected by "wiggles" in the decay curve. For the data shown in Fig. 6.8c, the fitted value of τ changes by a factor of nearly 2 with an incorrect choice of π-pulse duration.

6.4 Characterizing Linewidths (II): Sources of *Natural* (i.e., Inherent) Broadening

After minimizing or eliminating all experimental sources of broadening, what should be left are *natural* sources of broadening, i.e. physics influencing the lifetime of the excited state irrespective of the spectroscopy or trapping scheme. In the limit of zero perturbing fields, two effects dominate the lifetimes of 1_g states:

6.4 Characterizing Linewidths (II): Sources of *Natural* (i.e., Inherent) Broadening

predissociation and radiative decay. When magnetic fields are applied, these natural lifetimes can be modified dramatically through state mixing with nearby levels.

6.4.1 Radiative Decay

Radiative decay refers to the process by which an electronic excited state spontaneously decays to the ground state through emission of a photon. It can be described in terms of an interaction Hamiltonian $\hat{H}_{EM}$, representing the interaction of the electron cloud with a radiation field, having the following form [17]:

$$\hat{H}_{EM} = -\frac{iE_0 e}{m\omega} e^{i\vec{k}_\omega \cdot \hat{\vec{r}}} \vec{\epsilon} \cdot \hat{\mathbf{p}}, \tag{6.14}$$

where $\hat{\epsilon}$ is the electric field polarization unit vector, $\hat{\mathbf{r}}$ is the electronic position operator, $\hat{\mathbf{p}}$ is the electronic momentum operator, $\vec{k}$ is the EM wave vector, and I have omitted a sum over all electronic coordinates for clarity. The total decay rate can then be computed by evaluating the following transition element (i.e., by applying Fermi's golden rule):

$$\text{(Total transition rate)} \propto |\langle \Psi_1 | \hat{H}_{EM} | \Psi_2 \rangle|^2, \tag{6.15}$$

where $|\Psi_2\rangle$ represents a sum over all possible output channels.

However, it is much more common to Taylor-expand the interaction Hamiltonian to first or second order before computing, which is valid when $R/\lambda \ll 1$ and the radiation intensity is low (as is true for our experiment). A convenient classification groups these low-order terms into E1, M1, and E2 transitions:

$$\hat{H}_{EM} = -\frac{iE_0 e}{m\omega} \left[\underbrace{\vec{\epsilon} \cdot \hat{\mathbf{p}}}_{E1} + \underbrace{i(\vec{k}_\omega \cdot \hat{\mathbf{r}})(\vec{\epsilon} \cdot \hat{\mathbf{p}})}_{M1,E2} + \cdots \right]. \tag{6.16}$$

6.4.1.1 E1 Transitions

For molecules in *super*radiant states, electric dipole (E1) radiative decay to the electronic ground state is the dominant contributor to a state's natural lifetime. The E1 decay rate can be computed by calculating the transition dipole matrix for just the first term of Eq. (6.16):

$$\text{E1 decay rate} \propto \left| \langle \Psi_1 | \left(\frac{iE_0 e}{m\omega} \vec{\epsilon} \cdot \hat{\mathbf{p}} \right) | \Psi_2 \rangle \right|^2 \tag{6.17}$$

Using the well-known result $[\hat{\mathbf{r}}, \hat{H}_0] = \frac{i\hbar}{m}\hat{\mathbf{p}}$ [1] (where $\hat{H} = \hat{H}_0 + \hat{H}_{EM}$), we can rewrite the decay amplitude in a way more amenable to computation:

$$\langle\Psi_1|\left(\frac{iE_0e}{m\omega}\vec{\epsilon}\cdot\hat{\mathbf{p}}\right)|\Psi_2\rangle = \frac{iE_0e}{m\omega}\langle\Psi_1|\left(\frac{m}{i\hbar}\right)\vec{\epsilon}\cdot(\hat{\mathbf{r}}\hat{H}_0 - \hat{H}_0\hat{\mathbf{r}})|\Psi_2\rangle$$

$$= \frac{E_0e}{\hbar\omega}(E_2 - E_1)\langle\Psi_1|\vec{\epsilon}\cdot\hat{\mathbf{r}}|\Psi_2\rangle$$

$$= E_0e\langle\Psi_1|\vec{\epsilon}\cdot\hat{\mathbf{r}}|\Psi_2\rangle. \qquad (6.18)$$

Equation (6.18) contains no explicit dependence upon the size of the atom or bond length of the molecule in question, and so to a first approximation it's clear that the E1 radiative decay rate for rovibrational states in ^{88}Sr$_2$ should be independent of bond length. The story is different, however, for higher-order transitions.

6.4.1.2 M1/E2 Transitions

Subradiant states are defined as such because electric dipole radiation to the ground state is forbidden by parity selection rules. Higher-order radiation due to the second term of Eq. (6.16), however, is allowed. To understand the physics behind such transitions more clearly, we can make use of the following vector identity:

$$(\mathbf{A}\times\mathbf{B})\cdot(\mathbf{C}\times\mathbf{D}) = (\mathbf{A}\cdot\mathbf{C})(\mathbf{B}\cdot\mathbf{D}) - (\mathbf{B}\cdot\mathbf{C})(\mathbf{A}\cdot\mathbf{D}) \qquad (6.19)$$

to rewrite the "M1, E2" radiative term:

$$\frac{E_0e}{m\omega}\left[(\vec{\mathbf{k}}_\omega\cdot\hat{\mathbf{r}})(\vec{\epsilon}\cdot\hat{\mathbf{p}})\right] = \frac{E_0e}{m\omega}\left[(\vec{\epsilon}\cdot\hat{\mathbf{r}})(\vec{\mathbf{k}}_\omega\cdot\hat{\mathbf{p}}) + (\vec{\mathbf{k}}_\omega\times\vec{\epsilon})\cdot(\hat{\mathbf{r}}\times\hat{\mathbf{p}})\right] \qquad (6.20)$$

We can rewrite $\hat{\mathbf{r}}\times\hat{\mathbf{p}} = \hat{\mathbf{L}}$ and again make use of the identity $[\hat{\mathbf{r}}, \hat{H}_0] = \frac{i\hbar}{m}\hat{\mathbf{p}}$ to get:

$$\frac{E_0e}{m\omega}\left[(\vec{\mathbf{k}}_\omega\cdot\hat{\mathbf{r}})(\vec{\epsilon}\cdot\hat{\mathbf{p}})\right] = \frac{E_0e}{m\omega}\left[\frac{m}{i\hbar}(\vec{\epsilon}\cdot\hat{\mathbf{r}})(\vec{\mathbf{k}}_\omega\cdot(\hat{\mathbf{r}}\hat{H}_0 - \hat{H}_0\hat{\mathbf{r}})) + (\vec{\mathbf{k}}_\omega\times\vec{\epsilon})\cdot\hat{\mathbf{L}}\right]$$

$$(6.21)$$

$$= \frac{E_0e}{mc}\left[\underbrace{-im\omega_{12}(\vec{\epsilon}\cdot\hat{\mathbf{r}})(\vec{\mathbf{e}}_k\cdot\hat{\mathbf{r}})}_{E2} + \underbrace{(\vec{\mathbf{e}}_k\times\vec{\epsilon})\cdot\hat{\mathbf{L}}}_{M1}\right], \qquad (6.22)$$

where $\vec{\mathbf{e}}_k$ is a unit vector defining the radiated photon propagation direction, and I have again assumed a transition between states with energy difference $E_2 - E_1 = \hbar\omega_{12}$.

6.4 Characterizing Linewidths (II): Sources of *Natural* (i.e., Inherent) Broadening

By rewriting the above operators in the molecular frame, it's possible to see that both M1 and E2 decay rates should scale with the square of the bond length. For example, consider a diatomic molecule consisting of atoms A and B for which the internuclear axis is along z, so that x, y lay in the bisecting plane. Let atoms (A, B) be located at $(x, y, z) = \left(0, 0, \pm \frac{R}{2}\right)$. For M1 transitions, the transition moment due to the z-component of the $\hat{L}$ operator will be zero in the asymptotic, large-R limit. This can be seen by rewriting $\hat{L} = \hat{x}\hat{p}_y - \hat{y}\hat{p}_x$ and evaluating the transition moment using the properly symmetrized asymptotic wavefunctions [3, 15]. If we next consider just the x-component of the M1 operator (the y-component would give the same result), we can approximate the total angular momentum operator L_x as simply the sum of the individual atomic angular momentum operators [3]:

$$\hat{L}_x = \hat{L}_{xA} + \hat{L}_{xB} + \frac{R}{2}P_{yA} - \frac{R}{2}P_{yB} \tag{6.23}$$

Let the molecular wavefunction have the simplified form $|\Psi_{\text{mol}}\rangle = |\Psi_A\rangle|\Psi_B\rangle$. Matrix elements due to $\hat{L}_{xA}$ and $\hat{L}_{xB}$ will be zero since the atomic wavefunctions have zero angular momentum projection along axes perpendicular to the internuclear axis [3]. It is then straightforward to show that for transitions to molecular states in which one of the atom's energies has changed from E_1 to E_2, the M1 transition moment can be rewritten:

$$(\text{M1 transition moment}) \propto \langle \Psi_i | \hat{L}_x | \Psi_f \rangle \propto \omega_{12} R \langle \Psi_A^i | \hat{y}_A | \Psi_A^f \rangle \tag{6.24}$$

$$\rightarrow (\text{M1 decay rate}) \propto R^2 \tag{6.25}$$

The same result can be found for E2 transitions by performing a similar calculation. This time, only components of the E2 operator containing the z-axis (i.e., $(\vec{\epsilon} \cdot \hat{r}) = z$ or $(\vec{e}_k \cdot \hat{r}) = z$) will contribute to the transition moment, for reasons similar to those outlined above.

The implication of Eq. (6.25) is that, in the absence of other decay mechanisms, subradiant molecules with larger bond lengths (or, equivalently, smaller binding energies) should decay faster. Surprisingly, we discovered the exact opposite behavior: molecular lifetime for the subradiant states in our experiment *increased* with bond length (see, e.g., Fig. 6.5). Furthermore, the decay rates calculated with an ab initio molecular model gave decay rates on the order of 1 to ~several Hz, nearly three orders of magnitude smaller than the experimentally observed linewidths. Both of these facts support the idea that another source of inherent broadening must dominate the decay rate. And in fact, there is another well-known source of broadening of molecular lines not yet considered here: *predissociation*.

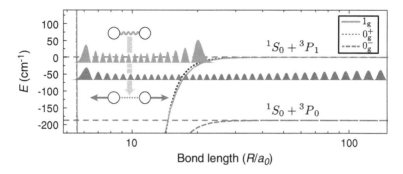

Fig. 6.9 A schematic representation of the predissociation process (adapted from [9])

6.4.2 Predissociation

Predissociation involves the direct tunneling of a bound molecular state to an unbound atomic state without the emission of a photon, and is represented schematically in Fig. 6.9. In order for energy to be conserved, the outgoing atomic fragments must carry away a large amount of kinetic energy. For ^{88}Sr$_2$ molecules in the 1_g potential, the dominant predissociative channel is to the 0_g^- threshold of $^1S_0+^3P_0$ molecules. We can estimate the amount of kinetic energy liberated in such an event as the energy difference between $^3P_0$ and $^3P_1$ atoms:

$$\begin{aligned}
\Delta E_{\text{pre}} &= E_{^3P_1} - E_{^3P_0} \\
&= 14504.351 \text{ cm}^{-1}(hc) - 14317.520 \text{ cm}^{-1}(hc) \\
&= 186.831 \text{ cm}^{-1}(hc) \\
&\approx 5.6 \text{ THz}(h)
\end{aligned} \quad (6.26)$$

The above value is *large*. For comparison, the most energetic atoms produced in the ultracold photodissociation studies described in this thesis (see Chap. 8) have energies of ~400 MHz. We can estimate the velocity of predissociated atom fragments using conservation of energy:

$$2 \times \frac{1}{2} m_{(^{88}Sr)} v^2 = 5.6 \text{ THz}(h)$$

$$\to v = \sqrt{\frac{5.6 \text{ THz}(h)}{m_{(^{88}Sr)}}}$$

$$\approx 159 \text{ m/s} \quad (6.27)$$

Since the entire field of view of our imaging system is <1 cm, atom fragments produced by predissociation will fly out of view in at most ~60 μs. This is

6.4 Characterizing Linewidths (II): Sources of *Natural* (i.e., Inherent) Broadening

unfortunately too fast to allow for direct imaging, since it's nearly twice as brief as the lifetime of even the shortest lived 1_g state described in this thesis. It is, however, fast enough to be confident that predissociated atom fragments will always be "invisible," meaning that unlike the possibility of spontaneous decay to the ground state (Sect. 6.3.2.1), this decay mechanism will not artificially influence our measured values for the lifetime.

Whereas radiative M1 and E2 decay rates *increase* with increasing bond length, predissociation in this case displays the opposite behavior. We can estimate the predissociative decay rate by again using Fermi's golden rule [9]:

$$\Gamma_{\text{pre}} \approx \frac{2\pi}{\hbar} |\langle 1_g, v', J', m'|\hat{H}_R|0_g^-, E, J', m'\rangle|^2, \tag{6.28}$$

where $|0_g^-, E, J', m'\rangle$ is the energy-normalized continuum wavefunction with energy E matching that of the bound level v'. $\hat{H}_R$ represents the coupling Hamiltonian between states of the 1_g and 0_g^- potentials, the physics of which is due to coupling between electronic angular momentum $\hat{j} = \hat{L} + \hat{S}$ and total angular momentum $\hat{J}$, and has the form $\hat{H}_R = -\frac{\hbar^2}{2\mu R^2}(\hat{J}_+\hat{j}_- + \hat{J}_-\hat{j}_+)$.

In the asymptotic, long-range limit, $\Gamma_{\text{pre}} = 0$ because $\langle 1_g|\hat{j}_\pm|0_g^-\rangle \approx \langle ^3P_1|\hat{j}_\pm|^3P_0\rangle = 0$ [9]. In the short-range limit, however, the matrix element is non-zero, and we can separate the transition element into radial and rotational parts:

$$\Gamma_{\text{pre}} \approx \frac{2\pi}{\hbar} |\langle \psi_{v'}(R)|\psi_f(E)\rangle|^2 \cdot f(J', m'), \tag{6.29}$$

where $\psi_{v'}(R)$ is the rotational part of the wavefunction describing the initial vibrational state, $\psi_f(E)$ is the final continuum wavefunction, and $f(J', m')$ is (constant with R) function of the angular momenta.

Evaluating Eq. (6.29) exactly would show that the predissociative decay rate is several orders of magnitude larger than the radiative decay rate for all subradiant states explored by our group except for $1_g(-1, 1)$, which possesses radiative and predissociative contributions of comparable magnitudes. This calculation, though, is beyond the scope of this thesis because it requires knowledge of the shapes of the initial and final wavefunctions (which our theorist collaborators in Poland have worked hard to provide). But we can get an intuitive understanding of the long-range behavior in the following way.

Taking a cue from quantum defect theory, we can write the rovibrational wavefunction in the following form [10]:

$$\Psi_v(R) = \left(\frac{\partial E_v}{\partial v}\right)^{1/2}_{E=E_v} \left(\frac{2\mu}{\pi\hbar^2}\right)^{1/2} \alpha_v(R, k) \sin(\beta_v(R, k)), \tag{6.30}$$

where $\frac{\partial E_v}{\partial v}$ is the vibrational spacing, $\alpha(R, k)$ and $\beta_v(R, k)$ are the quantum amplitude and phase, respectively, of the state v, and $k(R) = \sqrt{2[E_v - V(R)]\mu}/\hbar$ is the local wavenumber. Equation (6.30) is a convenient way to represent the vibrational wavefunction for the following reasons. First, since the rovibrational states we deal with are very weakly-bound compared to the depth of the potential, the functions $\alpha(R,k)$ and $\beta_v(R,k)$ are nearly constant at short range. Since the final continuum wavefunctions $|\psi_f(E)\rangle$ are also very nearly constant for all initial v' due to the large dissociation energy, it becomes clear that the predissociative linewidth will be a function only of the vibrational energy level spacing:

$$\Gamma_{\text{pre}}(v) = p \left(\frac{\partial E_v}{\partial v} \right)_{E=E_v}, \tag{6.31}$$

where p is a free parameter quantifying the overlap between the initial bound and final continuum wavefunctions.

We have direct experimental access to $\left(\frac{\partial E_v}{\partial v} \right)_{E=E_v}$ via numerical differentiation of the measured binding energies. We can also approximate it by applying the LeRoy-Berstein formula [7], which relates binding energy to vibrational number according to:

$$E_v = -[(v_D - v)f(n)]^{\frac{2n}{n-2}}, \tag{6.32}$$

where the potential is assumed to have the form $V(R) = -C_n/R^n$, v_D is the "effective" vibrational number of the dissociation limit, the vibrational energy E_v is defined as zero at the dissociation threshold, and $f(n)$ is a complicated function of n. Differentiating the above expression with respect to v yields:

$$\frac{\partial E_v}{\partial v} = \frac{2n}{n-2} f(n)[(v_D - v)f(n)]^{\frac{n+2}{n-2}} \tag{6.33}$$

$$\propto E_v^{\frac{n+2}{2n}} \tag{6.34}$$

$$\propto R^{-\frac{n+2}{2}}, \tag{6.35}$$

where the last equality came from equating a bound state's energy E_v with the value of the potential $V(R)$ at its bond length.

Equation (6.35) implies that the predissociation rate should scale differently in the large- and small-bond length regimes. At large length scales, the interaction potential for $^{88}\text{Sr}_2$ dimers can be expanded in terms of C_3 and C_6 coefficients [18] via $V(R) \approx -\frac{C_6}{R^6} + \frac{C_3}{R^3} +$ (higher order...). For relatively small bond lengths where the C_6 term dominates, we'd expect the decay rate $\Gamma_{\text{pre}} \propto R^{-\frac{6+2}{2}} = R^{-4}$, whereas at larger bond lengths we'd expect the decay to rate scale as $\Gamma_{\text{pre}} \propto R^{-\frac{3+2}{2}} = R^{-2.5}$. For rovibrational states within the 1_g potential, it turns out that the crossover occurs between the $1_g(-2, 1)$ and $1_g(-1, 1)$ states.

6.4 Characterizing Linewidths (II): Sources of *Natural* (i.e., Inherent) Broadening

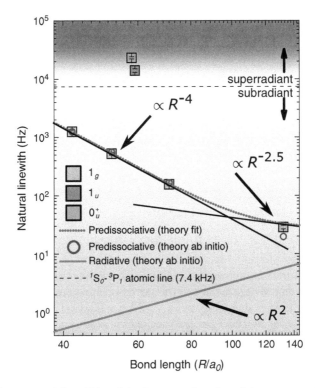

Fig. 6.10 The measured linewidths of the four most least-bound 1_g states are shown plotted against bond length, as well as those of two representative supperradiant states. The bond length is defined as the classical turning point for a state with binding energy E confined to its respective potential (1_u, 0_u, 1_g). Theory curves displaying radiative and predissociative contributions to the linewidth are shown in red and blue, respectively. Black lines show $\gamma \propto R^{-4}$ or $R^{-2.5}$ and serve to guide the eye. Note that while radiative decay rates *increase* with bond length, predissociative decay rates *decrease* with bond length (adapted from [9])

6.4.3 Results: Linewidth vs Bond Length at Zero Field

Figure 6.10 shows the measured values of linewidths for the four most least-bound 1_g states, as well as theory curves describing both the radiative and the predissociative contributions. Two superradiant state linewidths are also shown for comparison. In the figure, the theory curves are fits based on the approximations of Eqs. (6.31) and (6.35), while open circles denote ab initio calculations based on numerical solutions of Eq. (6.28) [9].

As is clear from the figure, the linewidths are completely dominated by the contribution from predissociation. Only for the narrowest, least-bound state, where the radiative linewidth is $\sim$30% of the predissociative linewidth, can the effects of radiative decay be observed.

Table 6.1 All values used for plotting data and theory curves in Fig. 6.10

ψ	E_b (MHz×h)	$\frac{\partial E}{\partial v}$ (MHz×h)	R (Bohr radii)	γ_{rad} (Hz)	γ_{pre} (Hz)	γ_{exp} (Hz)
$1_g(-1,1)$	19	132.3	132	5.7	19.7	28.5(2.0)
$1_g(-2,1)$	316	643.3	71	1.6	166	156.3(5.3)
$1_g(-3,1)$	1669	2244.3	52	0.8	555	525(30)
$1_g(-4,1)$	5168	4935.3	42	0.6	1243	1250(90)
$1_u(-1,1)$	353	N/A	59	N/A	N/A	14,000(1000)
$0_u^+(-4,1)$	1084	N/A	58	N/A	N/A	23,000(1000)

Calculations of predissociative and radiative decay linewidth contributions were performed by Wojciech Skomorowski and Robert Moszynski [9]

6.4.3.1 Details on Data Used in Fig. 6.10

Table 6.1 shows all theory and experiment values used to produce the data points and theory curves shown in Fig. 6.10.

Two theory curves are shown in Fig. 6.10, depicting linewidth contributions from predissociation and M1/E2 radiation. The predissociation theory curve is drawn to be proportional to the value for $\frac{\partial E}{\partial v}$ at all v, which was calculated numerically from knowledge of the binding energies E_b. Specifically, a third-order polynomial fit was fitted to a plot of E_b vs v, and then differentiated to obtain $\frac{\partial E_{b\,(poly-fit)}}{\partial v}|_{v=(-1,-2,-3,-4)}$. This fit had the following form:

$$E_{b\,(poly-fit)} = -312 - 711.\bar{3} \times v - 562 \times v^2 - 181.\bar{6} \times v^3, \quad (6.36)$$

The predissociative linewidths γ_{pre} were then calculated according to Eq. (6.31). The scaling coefficient p can in principle be calculated ab initio, but in our case was determined simply by fitting Eq. (6.31) to a plot of the experimentally determined linewidths γ_{exp} vs v for the three most deeply-bound states (ignoring the least-bound state because its linewidth is expected to reflect a significant contribution from radiative decay, rather than be dominated by predissociation). The value of p found by following this method was $p = 0.24806$.

In order to convert a plot of γ_{pre} vs v into a plot of γ_{pre} vs R, a function was found by trial and error which could approximate the correct value of $\frac{\partial E}{\partial v}$ at every R. It had the following form:

$$\frac{\partial E}{\partial v} \approx \exp[13.7616 - 0.16315 \times R + 1.0 \times 10^{-3} \times R^2 - 2.07185 \times 10^{-6} \times R^3] \quad (6.37)$$

This function gives results which differ from the calculated value for $\frac{\partial E}{\partial v}$ by no more than 3%, which is plenty accurate enough for a curve intended only to guide the eye. This curve was then multiplied by the scale factor p discussed in the previous paragraph in order to convert to linewidth, and is represented by the dotted blue curve in Fig. 6.10.

6.4 Characterizing Linewidths (II): Sources of *Natural* (i.e., Inherent) Broadening

The radiative decay curve (red line on Fig. 6.10) was found by fitting the calculated ab initio values for the radiative decay linewidth contributions (found by our collaborators Wojciech Skomorowski and Robert Moszynski) to the function $\gamma_{rad} = \kappa \times R^2$. The fitted value of κ was $3.26(3) \times 10^{-4}$ Hz/(Bohr radius)^2, and the residuals on the fit were small enough to be consistent with the difference being due to rounding error.

6.4.4 Magnetic Field Mixing of Nearby Levels

Somewhat surprisingly, we've found that the values we measure for linewidth can be strongly influenced by the application of small magnetic fields. This effect is unrelated to the spectroscopic blurring which can be caused by magnetic field gradients as discussed in Sect. 6.3.1.2, but rather represents a real modification of the excited state lifetime. Figure 6.11 illustrates this phenomenon, showing experimentally determined values for the linewidth of both $J = 1$ and $J = 2$ 1_g states (measured either "in the dark" or via spectroscopy depending on the magnitude of the linewidth).

Several interesting trends in this figure are worth pointing out. First, notice that while for the $J' = 1$ data depicted in Fig. 6.11a the linewidths *broaden* with increasing magnetic field, for the $J' = 2$ data depicted in Fig. 6.11b–e the linewidths actually *narrow*. Experimentally this is comforting because it proves that we are not simply observing an artificial, experimentally induced broadening mechanism, but rather are seeing something which reflects inherent molecular physics. Second, as implied by the dotted lines drawn to guide the eye, notice that the linewidths of all $1_g(v', J' = 1)$ states are well-described as increasing *quadratically* with magnetic field. One possible explanation for this behavior would be magnetic field-induced mixing of nearby states [8]. As was discussed in Chap. 5, small magnetic fields can induce a pure state $|\mu\rangle$ to become slightly mixed with its neighbors $|\nu\rangle$ according to

$$|\mu(B)\rangle \approx |\mu(0)\rangle + \sum_{\nu \neq \mu} (B/B_{\mu\nu})|\nu\rangle, \qquad (6.38)$$

where $B_{\mu\nu} = (E_\mu - E_\nu)/\langle\mu(0)|\hat{H}_{Zeeman}/B|\nu(0)\rangle$. In this case, radiative decay transitions between two states $|\gamma\rangle$ and $|\mu\rangle$ which were initially forbidden (i.e., $|\Omega_{\gamma\mu}|^2 = 0$) could become slightly allowed according to:

$$|\Omega_{\gamma\mu}(B)|^2 \approx B^2 \left| \sum_{\nu \neq \mu} \frac{\Omega_{\gamma\nu}(0)}{B_{\gamma\nu}} \right|^2, \qquad (6.39)$$

which would cause broadening that is quadratic with field, just as we observe.

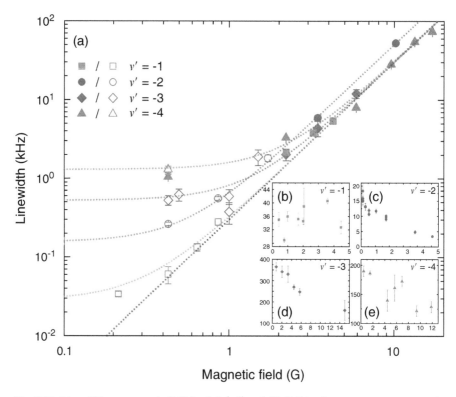

Fig. 6.11 Linewidths vs magnetic field for $1_g(v'; J' = 1, 2)$. Solid points represent spectroscopic measurements of linewidth, while hollow points represent "in the dark" measurements of lifetime. (**a**) Linewidths of the four least-bound $1_g(v', J' = 1)$ states plotted against magnetic field. The dotted lines are to guide the eye, and have the functional form $\gamma(B) = \gamma_0 + A \cdot B^2$, where γ_0 is the zero-field linewidth for each state as described in Table 6.1, and $A = 3 \times 10^{-4}$ Hz/G^2 for $v' = -1, -3, -4$ and $A = 5 \times 10^{-4}$ Hz/G^2 for $v' = -2$. (**b–e**) Linewidths of the four least-bound $1_g(v', J' = 2)$ states plotted against magnetic field. Note how for the majority of cases a clear *narrowing* of the linewidth is observed with increased magnetic field—the exact opposite of the behavior observed for $1_g(v', J' = 1)$ states

6.5 Comment on the Search for 0_g States

Over the course of my PhD, our group has devoted a significant amount of experiment time to the search for states within the 0_g potential. These searches consisted of creating a population in one of the $X(v = -1, -2; J = 0, 2)$ states, and then applying a strong, long-duration laser pulse which would deplete the population of the initial state if tuned into resonance with a 0_g state.

Unfortunately, each one of these searches has so far come up empty, despite laser sweeps across several GHz and theory guidance from our collaborators at the University of Warsaw. There are several possible reasons why our searches might have failed so far:

- **Transitions to 0_g states may simply be too weak to see with our setup.** The weakest transitions we've observed have been highly-forbidden, magnetic field-enabled transitions satisfying $\Delta J = 2, 3$ (see Fig. 5.4). Preliminary theory calculations have indicated that most of the transitions to 0_g states we hope to observe may be weaker than the weakest demonstrations we have so far demonstrated the capability to see. One possible way around this difficulty would be to prepare a spectroscopy scheme for which E1 transitions are allowed, such as a two-photon process starting from the metastable $1_g(-1, 1)$ state.
- **Predissociation (or another unknown mechanism) may broaden these states so much that losses due to the spectroscopy laser are difficult or impossible to disentangle from losses due to other effects.** For example, we sweep our lasers using acousto-optic modulators (AOMs) whose bandwidth is ~ 100 MHz. Since the amount of laser power passing through our AOMs depends upon frequency, shallow, broad dips in signal with widths comparable to or larger than the AOM bandwidth would be difficult to recognize as genuinely due to resonance with a broad transition to a 0_g state.
- **Transitions to 0_g states could be so narrow that we simply "stepped" over them.** For a given amount of time we want to invest in any spectroscopic search, we must make a choice between frequency step size and sweep range. We generally erred on the side of large sweep ranges by using step sizes of a few hundred kHz, reasoning that at large probe powers the transitions we hoped to find would be power-broadened enough to see. However, this assumption may not have been valid, and we may have stepped over 0_g states without realizing it.
- **The binding energies of 0_g states may be so close to those of 0_u and 1_u states that weak, E1-forbidden transitions to 0g states are effectively "hidden" by the strong, E1-allowed transitions to states of u-symmetry.** If this is true, then we could circumvent this problem by either operating at extremely low probe powers and very small step sizes (hoping that transitions to the 0_g states are both very narrow and spectroscopically resolvable from the ~ 15 kHz natural linewidth of transitions to the u-symmetry states), or by designing a spectroscopy scheme which severely suppresses transitions to u-symmetry states (such as the two-photon scheme starting from $1_g(-1, 1)$ described above).

We are still very interested in eventually finding these states, since there is the possibility that they may be even longer-lived than 1_g states, which would make them interesting from a metrological perspective. Perhaps the next generation of ZLab will discover this final piece of the puzzle and fill in the last gap in our understanding of weakly-bound strontium molecules.

References

1. Bethe, H., Salpeter, E.: Quantum Mechanics of One-and Two-Electron Atoms. Springer, Berlin (2012)
2. Bishof, M., Zhang, X., Martin, M., Ye, J.: Optical spectrum analyzer with quantum-limited noise floor. Phys. Rev. Lett. **111**(9), 093604 (2013)

3. Bussery-Honvault, B., Moszynski, R.: Ab initio potential energy curves, transition dipole moments and spin-orbit coupling matrix elements for the first twenty states of the calcium dimer. Mol. Phys. **104**(13–14), 2387–2402 (2006)
4. DeVoe, R., Brewer, R.: Observation of superradiant and subradiant spontaneous emission of two trapped ions. Phys. Rev. Lett. **76**(12), 2049 (1996)
5. Dicke, R.: Coherence in spontaneous radiation processes. Phys. Rev. **93**(1), 99 (1954)
6. Guerin, W., Araújo, M., Kaiser, R.: Subradiance in a large cloud of cold atoms. Phys. Rev. Lett. **116**, 083601 (2016)
7. LeRoy, R., Bernstein, R.: Dissociation energy and long-range potential of diatomic molecules from vibrational spacings of higher levels. J. Chem. Phys. **52**(8), 3869–3879 (1970)
8. McGuyer, B., McDonald, M., Iwata, G., Skomorowski, W., Moszynski, R., Zelevinsky, T.: Control of optical transitions with magnetic fields in weakly bound molecules. Phys. Rev. Lett. **115**(5), 053001 (2015)
9. McGuyer, B., McDonald, M., Iwata, G., Tarallo, M., Skomorowski, W., Moszynski, R., Zelevinsky, T.: Precise study of asymptotic physics with subradiant ultracold molecules. Nat. Phys. **11**(1), 32–36 (2015)
10. Mies, F.: A multichannel quantum defect analysis of diatomic predissociation and inelastic atomic scattering. J. Chem. Phys. **80**(6), 2514–2525 (1984)
11. Osborn, C.: The physics of ultracold Sr_2 molecules: optical production and precision measurement. Ph.D. Thesis (2014)
12. Porsev, S., Safronova, M., Clark, C.W.: Relativistic calculations of C_6 and C_8 coefficients for strontium dimers. Phys. Rev. A **90**(5), 052715 (2014)
13. Skomorowski, W., Moszynski, R., Koch, C.: Formation of deeply bound ultracold Sr_2 molecules by photoassociation near the $^1S + {}^3P_1$ intercombination line. Phys. Rev. A **85**(4), 043414 (2012)
14. Skomorowski, W., Pawłowski, F., Koch, C., Moszynski, R.: Rovibrational dynamics of the strontium molecule in the $A^1\Sigma_u^+$, $c^3\Pi_u$, and $a^3\Sigma_u^+$ manifold from state-of-the-art *ab initio* calculations. J. Chem. Phys. **136**(19), 194306 (2012)
15. Takasu, Y., Saito, Y., Takahashi, Y., Borkowski, M., Ciuryło, R., Julienne, P.: Controlled production of subradiant states of a diatomic molecule in an optical lattice. Phys. Rev. Lett. **108**(17), 173002 (2012)
16. Tojo, S., Hasuo, M.: Oscillator-strength enhancement of electric-dipole-forbidden transitions in evanescent light at total reflection. Phys. Rev. A **71**(1), 012508 (2005)
17. Tokmakoff, A.: Introductory Quantum Mechanics II, Spring 2009. Massachusetts Institute of Technology: MIT OpenCourseWare. Accessed 20 April 2016
18. Zelevinsky, T., Boyd, M., Ludlow, A., Ido, T., Ye, J., Ciuryło, R., Naidon, P., Julienne, P.: Narrow line photoassociation in an optical lattice. Phys. Rev. Lett. **96**(20), 203201 (2006)

Chapter 7
Carrier Thermometry in Optical Lattices

The first step of nearly every experiment in modern atomic physics consists of cooling an atomic or molecular sample. But why is this so universally important? Why can't accurate measurements be performed on room-temperature or "hot" samples?

The truth is that very accurate measurements of frequency *can* be made on hot samples, and that for many years (before the late 1980s) the best measurements of atomic properties were obtained with beam experiments using Ramsey's method of separated oscillating fields [12]. However, the accuracy of such devices suffered from a fundamental limitation. The linewidth γ of a transition under investigation with the Ramsey method is related to the time T the molecules are allowed to evolve between probe pulses, roughly according to the following equation:

$$\gamma \approx \frac{1}{T} \approx \frac{v}{d}, \tag{7.1}$$

where v is the velocity of the atoms under investigation and d is the length of the interaction region. Hotter atoms move faster, and so there are only two options for decreasing linewidth (and therefore improving frequency resolution): increase the length of the interaction region, or decrease the temperature of the atoms.

As a practical matter, the interaction region for a clock designed to fit within a room will be limited to no more than a few meters at most. And in fact, every primary NIST frequency standard from the late 1950s until 1998 included a very long pipe through which hot atoms could flow [7]. But with the invention of laser cooling in the late 1970s and early 1980s, a new possibility emerged. If atoms can be slowed to very small velocities, it becomes possible to envision performing Ramsey's separated fields technique on a pipe which has been oriented vertically, so that the time between interactions is simply the time it takes for an atom to ballistically fly up, reach the apex of its trajectory, and then fall down. For a pipe a few meters long, this results in an interaction time which has been increased from

© Springer International Publishing AG 2018
M. McDonald, *High Precision Optical Spectroscopy and Quantum State Selected Photodissociation of Ultracold 88Sr2 Molecules in an Optical Lattice*, Springer Theses, https://doi.org/10.1007/978-3-319-68735-3_7

tens of ms to as much as a second, with a corresponding decrease in linewidth and increase in frequency resolution. The current US frequency standard, NIST-F2, is based on this technique [11].

In order to achieve even longer interrogation times, not limited by the free-fall of an atom over a few-meter-long path, the most accurate clocks today make use of optical lattices or ion traps [4, 10]. The advantage of this technique is an improvement of both the duration and the ability to control the interrogation time. The drawback is the introduction of new systematic shifts due to the presence of the trapping potential, which must be controlled and corrected for. In our experiment, as well as other optical lattice clock experiments, we operate a so-called "magic wavelength lattice," whereby the lattice wavelength is tuned so that the trap depth is equivalent for atoms or molecules in both the initial and final states of the transition [5]. In such a trap, if the wavelength of the optical lattice is at all non-magic, then the "blurring out" of the lineshape of the transition under investigation will be influenced by the temperature of the atoms or molecules within the trap. It is therefore of great interest to know *exactly* the temperature of the trapped atoms or molecules.

Today, measuring and lowering ultracold temperatures is an active field of research in modern optical lattice experiments [9], and many different approaches exist. For thermal clouds of atoms, the most common technique for determining temperature is "time of flight imaging," whereby a cloud of atoms is released and their locations are recorded after having been allowed to expand for some amount of time [6]. Since hotter atoms move faster, the temperature of the cloud can be inferred from its spatial extent. This method, however, relies on the ability to image the particles undergoing expansion. For atoms this is possible because atoms generally possess strong cycling transitions, which allow for the possibility of absorption imaging. For molecules, however, there is generally no easy way to determine their temperature.

In this chapter I will describe a new technique for determining the temperature of atoms or molecules confined to an approximately harmonic trap, which depends on accurately recording the lineshape of a transition "smeared out" by the thermal distribution of particles in a slightly non-magic trap. Our technique offers a solution for determining temperature when narrow transitions are available (in any frequency regime). This chapter represents a more detailed analysis of work first published by myself and colleagues in 2015 [8].

7.1 A Roadmap for Determining Temperature

In order to determine the temperature of a sample of ∼harmonically-trapped particles, it is necessary to measure only three "frequencies": (1) the differential trap depth for the transition, manifest as a non-zero light shift; (2) the axial trap spacing; and (3) the full width at half maximum (FWHM) of the thermally-broadened lineshape.

7.1 A Roadmap for Determining Temperature

Each measurement requires some care, and it will take several pages to derive the exact relationships between all of these quantities. But when traveling through a dense forest it's often helpful to have a compass and a map, so as to have some idea about where we hope to end up. So with that in mind, here is a roadmap (justified and elaborated upon in the following sections) describing the sequence of measurements which must be made in order to determine the temperature of trapped particles:

1. Under *non-magic* ($\alpha'/\alpha \neq 1$, where α and α' are the initial and final trap polarizabilities, respectively) conditions, record the carrier spectrum at several lattice powers and fit with the following lineshape:

$$N(f) = \begin{cases} N_0 \exp\left[A(f-f_0)^2 e^{-B(f-f_0)}\right], & B(f-f_0) \geq 0 \\ N_0, & B(f-f_0) < 0 \end{cases} \quad \text{(Eq. (7.28))}$$

2. Plot the fitted values of f_0 vs lattice power P according to the following equation:

$$f_0 = L_0 P \equiv \frac{1}{h}\left(1 - \frac{\alpha'}{\alpha}\right) U_0 \quad \text{(Eq. (7.50))}$$

3. Under *magic* ($\alpha'/\alpha = 1$) conditions, record the axial trap frequency f_x vs lattice power P according to the following equation:

$$f_x \equiv \left(\frac{\omega_x}{2\pi}\right) = \kappa P^{\frac{1}{2}} \quad \text{(Eq. (7.52))}$$

The trap frequency can be determined several ways. In this thesis we take high resolution spectra of the red and blue sidebands, fitted with the following equation:

$$N(f) = \begin{cases} N_0 \exp\left[C(f-f_\pm)^3 e^{-D(f-f_\pm)}\right], & D(f-f_\pm) \geq 0 \\ N_0, & D(f-f_\pm) < 0, \end{cases} \quad \text{(Eq. (7.43))}$$

and determine the trap spacing as $f_x = \frac{1}{2}(f_+ - f_-)$.

4. With knowledge of the light shift L_0 and the axial trap frequency f_x, the temperature T_{carrier} can be extracted from the fits of the *non-magic* carrier lineshapes, using the following equation:

$$T_{\text{carrier}} = h\left[k_B B\left(\sqrt{1 - \frac{2hL_0}{\lambda^2 \kappa^2 M}} - 1\right)\right]^{-1} \approx \frac{\lambda^2 \kappa^2 M}{k_B B L_0} \quad \text{(Eq. (7.56))}$$

The details of why the above equations are valid, and how measurements can be faithfully obtained, will be explained in the following sections.

7.2 Overview: Setting Up the Problem

The advantage of performing spectroscopy in a 1D optical lattice is that motion along the axial (x) direction is strongly quantized. This fact allows for the difference in initial and final motional trap states along the axial direction to be selected for by changing the frequency of the probe laser. Transitions which conserve motion along the axial direction are called *carrier transitions*. Transitions which add or subtract an axial motional trap quantum are called *sideband transitions*.

The popular method traditionally used for determining the temperature of trapped particles is to study the structure of their sideband transitions. Transitions which *add* a motional trap quantum are called *blue sidebands*, while transitions which *subtract* a motional trap quantum are called *red sidebands*. A cartoon illustrating this process is shown in Fig. 7.1a, while an example ^{88}Sr$_2$ spectrum showing red and blue sidebands is shown in Fig. 7.1b.

Because molecules which are already in the lowest possible trap state cannot be cooled further, the red sideband in a spectrum will always be smaller than the blue sideband. This fact is commonly taken advantage of in order to estimate the temperature of a sample. Since the motional states will be occupied with probabilities determined by the Boltzmann distribution, the ratio of the integrated sideband absorption cross sections will obey the following equation [1]:

$$\frac{\sigma_{\text{red}}^{\text{total}}}{\sigma_{\text{blue}}^{\text{total}}} = \frac{\sum_{n_x=1}^{N_x} e^{-E_{n_x}/(k_B T_x)}}{\sum_{n_x=0}^{N_x} e^{-E_{n_x}/(k_B T_x)}} = 1 - \frac{e^{-E_0/(k_B T_x)}}{\sum_{n_x=0}^{N_x} e^{-E_{n_x}/(k_B T_x)}}, \quad (7.2)$$

where temperature is labelled T_x to be clear that this is an *axial* temperature, i.e. it describes only the probability distribution of axial trap states while ignoring the

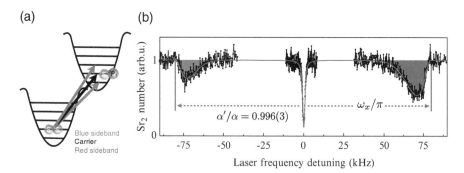

Fig. 7.1 (a) An illustration of a molecule in the electronic ground state undergoing a transition to the electronic excited state via a red sideband ($\Delta n_x = -1$), carrier ($\Delta n_x = 0$), or blue sideband ($\Delta n_x = +1$) transition. (b) A spectrum (adapted from [8]) showing sideband and carrier transitions under nearly-magic lattice conditions. The spectrum depicts a narrow ($\lesssim 30$ Hz) transition from a stable ground state to a metastable subradiant state, $X(-1, 0) \rightarrow 1_g(-1, 1)$. The sidebands were exposed for $\sim 60\times$ longer than was the carrier in order to produce a significant signal

7.2 Overview: Setting Up the Problem

radial trap state distribution. For a perfectly harmonic trap, $E_{n_x} = \hbar\omega_x(n_x + \frac{1}{2})$ and $N_x \to \infty$, and the equation above can be solved for T_x:

$$T_x^{(\text{harmonic})} = \frac{\hbar\omega_x}{k \ln\left(\frac{\sigma_{\text{red}}^{\text{total}}}{\sigma_{\text{blue}}^{\text{total}}}\right)} \tag{7.3}$$

In reality, our trap is not exactly harmonic, and in fact N_x will be finite. With a trap spacing of ~70 kHz (trap depth of ~50 μK), there should exist ~ $\frac{U_0}{\hbar\omega_x} \approx 15$ axial trap quanta. But given a temperature of ~5 μK, over 90% of the molecules will occupy just the first four axial trap states, implying that the harmonic approximation of Eq. (7.3) isn't so bad.

There are, however, a few flaws in defining temperature in this manner. Besides the fact that this method for measuring temperature is only sensitive to one dimension, it is also frequently difficult to measure with high confidence the ratio of areas between two sidebands, particularly at low temperatures where the red sideband becomes vanishingly small. These facts serve as motivation to find another technique for reliably measuring temperature.

7.2.1 Modeling the Trap Potential

The trap potential is formed by a retro-reflected Gaussian laser beam. Ignoring the (small) perturbation due to gravity, this potential can be modeled with the following equation [2, 3, 8]:

$$U(\vec{r}) = -U_0 e^{-2r^2/w(x)^2} \cos^2(2\pi x/\lambda), \tag{7.4}$$

where $w(x)$ is the lattice waist at position x with a minimum value of w_0, and λ is the wavelength of the light used to form the trapping potential. U_0 is the trap depth, given in terms of the total trapping laser power P by

$$U_0 = 4\alpha P/(\pi w_0^2 c\epsilon_0), \tag{7.5}$$

where α is the *polarizability* of the molecules in the trap. (Note that this polarizability will generally depend upon the rovibrational state of the molecule being trapped.)

To simplify the analysis of this system, we note that near the lowest energy states of this trap the potential will appear to be approximately harmonic [2]. With this approximation, we can rewrite the trapping potential near the center of the well (i.e., when $w(x) = w_0$) in the following way:

$$U(\vec{r}) \approx \frac{1}{2}M\omega_x^2 x^2 + \frac{1}{2}M\omega_r^2 r^2 - U_0, \tag{7.6}$$

where the axial and radial trap frequencies (ω_x and ω_r, respectively) have the following forms:

$$\omega_x = (2\pi/\lambda)\sqrt{2U_0/M} \tag{7.7}$$

$$\omega_r = (2/w_0)\sqrt{U_0/M} \tag{7.8}$$

where M is the mass of the trapped molecule.

Assuming that particles are trapped near the optical lattice focus (i.e., that the trapping potential is symmetric about the origin), the next highest order corrections to the harmonic approximation above would be quartic, and would have the following form:

$$V_{xx}(\mathbf{r}) = -\left(M\omega_x^2 x^2/2\right)^2/(3U_0) \tag{7.9}$$

$$V_{xr}(\mathbf{r}) = -\left(M\omega_x^2 x^2/2\right)\left(M\omega_r^2 r^2/2\right)/U_0 \tag{7.10}$$

$$V_{rr}(\mathbf{r}) = -\left(M\omega_r^2 r^2/2\right)^2/(2U_0) \tag{7.11}$$

While it turns out that these fourth-order corrections will have precisely no effect on our derivation for the lineshape of a non-magic carrier transition (which incidentally is one of the reasons this method for determining temperature is so powerful), they will be critically important for the derivation of sideband transitions.

7.2.1.1 Visualizing the Harmonic Approximation

We can get an intuitive picture of how temperature can affect lineshape by first picturing the limiting scenario of a perfectly harmonic, magic-wavelength trap in one dimension, with motional states separated by $\hbar\omega$. In this scenario, the separation between adjacent trap states is constant no matter how high up in the potential one goes. It is then easy to see that the lineshape for both carrier and sidebands should consist of a very narrow peak, since the same laser frequency is required to drive a transition no matter which motional trap state a particle initially occupies. Figure 7.2a illustrates this scenario.

If the trap is not exactly harmonic, however, then the separation between adjacent trap states is no longer constant at higher trap energies. The laser frequency required to drive a sideband transition therefore changes as the molecule's initial motional energy increases, resulting in a sideband lineshape which becomes smeared out. The carrier lineshape, however, remains a narrow peak, since it is insensitive to differences between adjacent levels. This scenario is depicted in Fig. 7.2b.

Finally, if the trap is both anharmonic *and* non-magic, then both carrier and sideband transitions will become blurred. The blurring of the carrier is caused by the difference in trap depths, which creates a progressively worse mismatch among initial and final motional state energies as the initial trap state energy increases. This scenario is depicted in Fig. 7.2c.

7.2 Overview: Setting Up the Problem

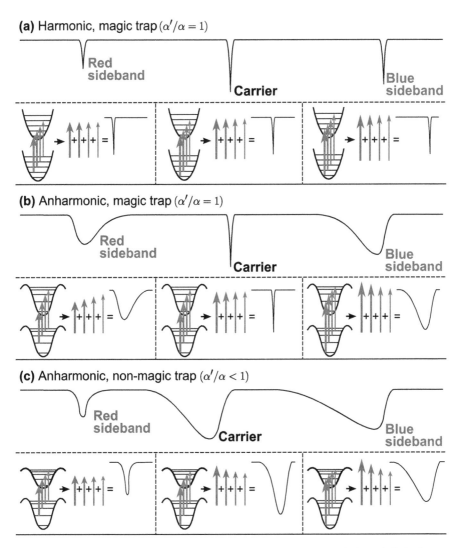

Fig. 7.2 In the illustrations above, an arrow represents a transition from $|n_x\rangle \to |n'_x\rangle$, its length represents the frequency of this transition, and its thickness represents the occupation number of the initial state $|n_x\rangle$. (**a**) For a perfectly harmonic, magic trap, the lineshapes describing all transitions will consist of narrow peaks, since the frequencies of both carrier and sideband transitions are independent of the initial state. (**b**) When a magic trap is anharmonic, sideband transitions become "blurred out" due to the variable spacing between axial trap state energies. The carrier, however, remains a narrow peak. (**c**) If the trap is non-magic, both carrier and sidebands will be subject to broadening. Notice, however, that in certain cases "broadening" due to the non-magic nature of the trap can be counteracted by "narrowing" due to the trap's anharmonicity, as is illustrated in the red sideband above

7.2.2 Calculating Lineshapes

To determine the lineshape of a transition will require summing over transition probabilities for many molecules distributed across a wide range of motional trap states in the lattice. But to begin, let's consider just a single particle of mass M confined to a single well of an optical lattice, and ask about the difference in energies between initial state $|\Psi_i\rangle$ and final state $|\Psi_f\rangle$. The evolution of its initial and final wavefunctions will be governed by the Hamiltonians

$$H_i = -\frac{\hbar^2}{2M}\nabla^2 + U(\vec{r}) + E_i \text{ and}$$
$$H_f = -\frac{\hbar^2}{2M}\nabla^2 + U'(\vec{r}) + E_f, \quad (7.12)$$

where $U(\vec{r})$ and $U'(\vec{r})$ describe the initial and final lattice potentials, and E_i and E_f describe the initial and final internal energies (i.e., the lattice-free energies), respectively.

The eigenfunctions of Eq. (7.12) will be harmonic oscillator states which can be characterized by three numbers (n_x, n_y, n_z) describing which motional states are occupied. Let's consider the following transition:

$$|\Psi_i\rangle = |n_x n_y n_z\rangle \rightarrow |\Psi_f\rangle = |n'_x n'_y n'_z\rangle \quad (7.13)$$

We can characterize a transition between initial and final state generically with the condition $n'_i = n_i + D$. *Carrier transitions* are defined as those preserving motional trap state, i.e. $D = 0$, while *sideband transitions* change the motional trap state by 1, i.e. $D = \pm 1$. Whether or not this condition is satisfied depends on how the experiment is constructed.

In our experiment, we interrogate molecules confined to a 1D optical lattice in the axial *resolved sideband regime* (RSB). This means that the linewidths of the transitions we study ($\sim$100 Hz $\rightarrow$ $\sim$20 kHz) are smaller than the axial trap spacing ($\sim$60 kHz), allowing us to drive either axial *carrier* or axial *sideband* transitions simply by adjusting the frequency of the probe laser. I will also assume that *radial sideband transitions* are driven at a negligible rate, i.e. that $(n_x, n_y) = (n'_x, n'_y)$. This is true for our experiment because we operate in the Lamb-Dicke regime and with a very small mismatch between the probe direction and 1D lattice axis [1].

The energy of a molecule in the $|n_x n_y n_z\rangle$ state is given by the following expression:

$$E_{n_x n_y n_z} = \hbar\omega_x\left(n_x + \frac{1}{2}\right) + \hbar\omega_y\left(n_y + \frac{1}{2}\right) + \hbar\omega_z\left(n_z + \frac{1}{2}\right)$$
$$= \hbar\omega_x\left(n_x + \frac{1}{2}\right) + \hbar\omega_r(n_y + n_z + 1), \quad (7.14)$$

7.2 Overview: Setting Up the Problem

where the second equality is due to the cylindrical symmetry of the potential (i.e., $\omega_y = \omega_z \equiv \omega_r$) and I have ignored the energy E_i due to the internal degrees of freedom of the molecule. The lineshape can be calculated by performing a weighted sum over frequencies required to drive all possible transitions from $|n_x n_y n_z\rangle$ to $|n'_x n'_y n'_z\rangle$.

The required driving frequencies will simply equal the energy differences δE between these states, which can be calculated according to the following equation:

$$\delta E \equiv \langle n'_x n'_y n'_z | H_f | n'_x n'_y n'_z \rangle - \langle n_x n_y n_z | H_i | n_x n_y n_z \rangle \qquad (7.15)$$

7.2.2.1 Carrier Transitions in a Non-magic Lattice ($\alpha'/\alpha \neq 1$)

Using the harmonic approximation (Eq. (7.14)), it's easy to show that for carrier transitions ($n_i = n'_i$) the energy difference can be divided into two components:

$$\delta E = \delta E_x + \delta E_r = \hbar(\omega'_x - \omega_x)\left(n_x + \frac{1}{2}\right) + \hbar(\omega'_r - \omega_r)(n_r + 1). \qquad (7.16)$$

Note that though the occupation number n_i is preserved in the transition, the trap spacing ω_i will generally change. We can rewrite the above equation in a more convenient form by realizing that since $\omega_i \propto \sqrt{U_0}$ and $U_0 \propto \alpha$, then:

$$\delta E_x = (\sqrt{\alpha'/\alpha} - 1)\hbar\omega_x\left(n_x + \frac{1}{2}\right)$$
$$\delta E_r = (\sqrt{\alpha'/\alpha} - 1)\hbar\omega_r(n_r + 1) \qquad (7.17)$$

These equations describe the change in energy accumulated by a particle undergoing a carrier transition between simple harmonic oscillator eigenstates. To obtain the total lineshape, we must now evaluate the weighting factor for a particular δE, and then sum over all possibilities. To proceed, let's consider δE_x and δE_r separately.

Probability Distribution for δE_x The x-direction is relatively straightforward, since the degeneracy for all axial trap states is 1. The weighting factor governing how strongly-represented a δE-transition will be should then just equal the occupation probability for the initial state, which itself is simply governed by the Boltzmann distribution, i.e.:

$$p_x(\delta E_x) = p_x(n_x) = \frac{1}{Z_x} e^{u(\delta E_x)}, \qquad (7.18)$$

where $Z_i = \sum_0^\infty e^{\frac{\hbar\omega_i}{k_B T}(n_i + \frac{1}{2})} = \frac{1}{2}\mathrm{csch}\left(\frac{\hbar\omega_i}{2k_B T}\right)$ and $u(\delta E_i) = \frac{\delta E_i}{k_B T(\sqrt{\alpha'/\alpha} - 1)} \geq 0$ is defined for clarity.

Note that this represents a discrete probability distribution parameterized by the Boltzmann step size of $\Delta_i = \hbar\omega_i/(k_B T)$.

Probability Distribution for δE_r The radial direction is slightly trickier, since the energy degeneracy $g(n_y, n_z)$ of the $|n_y, n_z\rangle$ state is not simply equal to 1, but rather:

$$g(n_y, n_z) = n_y + n_z + 1. \tag{7.19}$$

But not *so* tricky. The probability for a transition to be characterized by a particular δE_r is now simply the Boltzmann probability characterizing the occupation of a state $|n_x, n_y\rangle$, i.e. $p(n_y, n_z) = p(n_y)p(n_z)$, multiplied by the energy degeneracy $g(n_y, n_z)$:

$$\begin{aligned}
p_r(\delta E_r) &= g(n_y, n_z) p(n_y) p(n_z) \\
&= (n_x + n_y + 1) \left(\frac{1}{Z_y} e^{u(\delta E_y)}\right) \left(\frac{1}{Z_z} e^{u(\delta E_z)}\right) \\
&= \frac{1}{Z_r^2} \frac{1}{\Delta_r} u(\delta E_r) e^{u(\delta E_r)},
\end{aligned} \tag{7.20}$$

where $u(\delta E_i)$, Δ_i, and Z_i are defined as before.

Probability Distribution for δE The probability that a particular energy difference δE will be represented in our lineshape is then just equal to the probability that a state characterized by $|n_x, n_r\rangle$, satisfying $\delta E_x + \delta E_r = \delta E$, will be represented. To calculate the exact, discrete lineshape, we should then evaluate the following sum:

$$p_{\text{discrete}}(\delta E) = \sum_{\{n_x, n_r\}_{\delta E}} p(\delta E_x) p(\delta E_r) \tag{7.21}$$

where the set "$\{n_x, n_r\}_{\delta E}$" denotes all pairs (n_x, n_r) satisfying the condition $\delta E_x(n_x) + \delta E_r(n_r) = \delta E$.

To obtain a more convenient analytical form for this sum, we can make the approximation of converting the discrete sum into a continuous integral. For the radial coordinate this is clearly an excellent approximation. The radial trap frequency for our experiment is ~600 Hz, while the temperature is ~5 µK, and so $\Delta_r \approx 0.006 \ll 1$. For the axial coordinate, the step size $\Delta_x \approx \frac{70\,\text{kHz}}{5\,\mu\text{K}} \approx 0.7$. This also turns out to be small enough to reasonably approximate the discrete sum as a continuous integral—see Fig. 7.3.

In the continuum limit, the probability density should have the form:

$$\bar{p}_i[u(\delta E_i)] = \lim_{\Delta_i \to 0} p_i(\delta E_i)/\Delta_i \tag{7.22}$$

Using $\lim_{x \to 0} \operatorname{csch}(x) = \frac{1}{x}$, we find:

$$\bar{p}_x[u(\delta E_x)] = \begin{cases} e^{-u(\delta E_x)}, & u \geq 0 \\ 0, & u < 0 \end{cases} \tag{7.23}$$

7.2 Overview: Setting Up the Problem

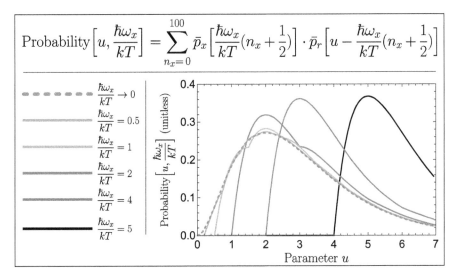

$$\text{Probability}\left[u, \frac{\hbar\omega_x}{kT}\right] = \sum_{n_x=0}^{100} \bar{p}_x\left[\frac{\hbar\omega_x}{kT}\left(n_x + \frac{1}{2}\right)\right] \cdot \bar{p}_r\left[u - \frac{\hbar\omega_x}{kT}\left(n_x + \frac{1}{2}\right)\right]$$

Fig. 7.3 Computations of the discrete lineshape given by Eq. (7.21), with various values for $\Delta_x = \frac{\hbar\omega_x}{k_B T}$. The dashed blue curve represents the limit of $\Delta_x \to 0$. Curves were plotted in Mathematica. The case of Δ_x as large as 1, though noticeably lumpy, still hews closely to the continuum limit (courtesy of Dr. Bart H. McGuyer)

$$\bar{p}_r[u(\delta E_r)] = \begin{cases} u(\delta E_r)e^{-u(\delta E_r)}, & u \geq 0 \\ 0, & u < 0 \end{cases} \quad (7.24)$$

The final expression for the probability density $\bar{p}[u(\delta E)]$ will then be given by the following convolution:

$$\bar{p}[u(\delta E)] = \int_0^\infty \bar{p}_r(u - u_x)\bar{p}_x(u_x)du_x = \begin{cases} \frac{1}{2}u^2 e^{-u}, & u \geq 0 \\ 0, & u < 0 \end{cases} \quad (7.25)$$

This Is the Result We've Been Looking For It describes the probability density for a given value of δE to be represented among energy differences between thermally distributed initial and final simple harmonic oscillator states. Equivalently, it is the strength at which a laser will be absorbed by a sample of molecules in a thermally distributed ensemble of simple harmonic oscillator states, assuming that the strength of a particular transition from $|n_x n_y n_z\rangle \to |n'_x n'_y n'_z\rangle$ is independent of (n_x, n_y, n_z). This assumption is reasonable so long as we are operating in the Lamb-Dicke regime and α'/α isn't too far from unity, which is true for the carrier transition data presented in this thesis. However, we will later see that this assumption is *not* valid for the case of sideband transitions.

Comment: The Effect of Anharmonic Corrections to the Potential Equation (7.25) was derived under the assumption that the trapping potential is perfectly harmonic. Of course this assumption is not perfectly satisfied, and so we might wonder how strongly our result will change as a result of the trap's deviation from harmonicity.

As was discussed in Sect. 7.2.1, near the center of a symmetric trap the leading order corrections to the harmonic potential will be quartic (Eqs. (7.9)–(7.11)). To evaluate whether these corrections will induce broadening, we should consider the first-order correction δE_{ij} to the original δE of the following form:

$$\delta E_{ij} \approx \langle n'_x n'_y n'_z | V_{ij} | n'_x n'_y n'_z \rangle - \langle n_x n_y n_z | V_{ij} | n_x n_y n_z \rangle \tag{7.26}$$

We don't yet need to explicitly evaluate these expressions, however. Using $\langle n_x | \hat{x}^2 | n_x \rangle = \frac{\hbar}{2M\omega_x}(2n_x + 1)$ and $\langle n_x | \hat{x}^4 | n_x \rangle = \frac{3\hbar^2}{2M^2\omega_x^2}(n_x^2 + n_x + \frac{1}{2})$, it is straightforward to see that:

$$\langle n_x n_y n_z | V_{ij} | n_x n_y n_z \rangle \propto \frac{\hbar^2 \omega_i \omega_j}{U_0} \times (\text{A function of } n_x, n_r) \tag{7.27}$$

Notice that since $\omega_x, \omega_r \propto \sqrt{U_0}$, the expression $\langle n_x n_y n_z | V_{ij} | n_x n_y n_z \rangle$ can be re-written in a way which is clearly *independent of the trap depth* U_0. Therefore for carrier transitions, where $(n'_x, n'_r) = (n_x, n_r)$, we will have $\delta E_{ij} = 0$, which means that these corrections *cannot contribute broadening to the carrier lineshape*.

This result is important. It means that the lineshape we derived, which is exact for harmonic traps, is also an extremely good approximation even when fourth-order corrections to the potential are incorporated. Note however that for sideband transitions ($n'_x = n_x + D$) the same result does not apply: we will need to account for these anharmonic corrections when deriving the full lineshape.

A Practical Carrier Lineshape for Fitting Equation (7.25) can be fitted to a spectroscopic lineshape as is shown later in Fig. 7.6g. In practice, we describe the transition probability with the following equation:

$$\bar{p}(f) = A(f - f_0)^2 e^{-B(f-f_0)} \tag{7.28}$$

where A is a catch-all scaling factor, f_0 is the zero-temperature transition frequency, and $B = \frac{h}{k_B T(\sqrt{\alpha'/\alpha} - 1)}$. And in fact, we don't *actually* measure a probability, but rather the number of molecules remaining after a probe laser has been shone onto a molecular sample. Similar to the discussion in Sect. 3.3.1, we can write a differential equation describing the rate at which molecular transitions are observed:

$$\frac{dN}{dt} = -\bar{p}(f) \cdot N, \tag{7.29}$$

7.2 Overview: Setting Up the Problem

with the following solution:

$$N(f) = N_0 \exp\left[-A(f-f_0)^2 e^{-B(f-f_0)}\right], \qquad (7.30)$$

where I have folded the probe time τ into the overall scaling factor A. Equation (7.30) is what's fitted to real data in our experiment. The value of "B" can be determined from the fit, but extracting the temperature T requires a separate measurement of the polarizability ratio. We can do this without having to rely on knowledge of the probe waist w_0 by combining measurements of the total light shift and the axial trap spacing. To measure the axial trap spacing accurately, however, requires a better understanding of the shape of axial sideband transitions.

7.2.2.2 Sideband Transitions When $(\alpha/\alpha') = 1$

The exact sideband lineshape can be calculated numerically from the shape of the lattice potential and resulting energies of the radial and axial trap states [1]. However, an analytical solution can also be obtained by approximating the exact lattice potential shape as harmonic plus lowest order corrections.

Let's consider the lineshape for sideband transitions in a perfectly magic lattice, i.e. where $(\alpha'/\alpha) = 1$. From the qualitative discussion of Fig. 7.2, it's clear that blurring in the lineshape will be entirely due to deviations from perfect trap harmonicity. Therefore we need only consider energy differences δE_{SB} incorporating higher order corrections (Eqs. (7.9)–(7.11)) to the harmonic potential (Eq. (7.6)) in order to derive a lineshape:

$$\delta E_{SB} = \delta E_{xx} + \delta E_{xr} + \delta E_{rr}. \qquad (7.31)$$

These new terms δE_{xx}, δE_{xr}, and δE_{rr} can be calculated according to Eq. (7.15), where δE_{ij} considers only the portion of the Hamiltonian H defined by the perturbation V_{ij}.

From the discussion of the previous section, we also know that broadening is contributed only by transitions which change the motional trap state. Since we are assuming only axial sideband transitions ($n'_x = n_x + D$ and $n'_r = n_r$), it's clear that $\delta E_{rr} = 0$ for a perfectly magic lattice, and therefore only terms resulting from a change in the axial trap number (δE_{xx} and δE_{xr}) will contribute to the final lineshape. We can then proceed as before, so long as we incorporate another bit of physics.

Inhomogeneous Excitation of Axial Sidebands Unlike carrier transitions, the strength of an axial sideband transition *does* in fact depend upon the initial occupation number n_x. This dependence is governed by the following equation:

$$\Omega(n_x, D)^2 \propto |\langle n'_x = n_x + D | e^{ik\hat{x}} | n_x \rangle|^2 \approx \begin{cases} 1 & D = 0 \\ \eta^2 \, n_x & D = -1 \\ \eta^2 \, (n_x + 1) & D = +1, \end{cases} \qquad (7.32)$$

where the LD parameter $\eta = k\sqrt{\hbar/(2M\omega_x)}$ and the axial wavenumber $k = 2\pi/\lambda$. *(This result can be derived by expanding $e^{ik\hat{x}} \approx 1 + ik\hat{x}$ and rewriting $\hat{x}$ in terms of creation and annihilation operators.)*

To calculate the final lineshape contribution from each term, we must (1) determine the dependence of each term upon (n_x, n_r), (2) write down the occupation probability for the n_i present in that term and participating in the transition, and (3) multiply by the transition strength given by Eq. (7.32). Let's again consider each term individually.

Probability Distribution for δE_{xr} Using the results from the discussion preceding Eq. (7.27), we can calculate the following expression:

$$\langle n_x n_y n_z | V_{xr} | n_x n_y n_z \rangle = -\frac{\hbar\omega_x \hbar\omega_r}{4U_0}(n_x + 1/2)(n_r + 1) \tag{7.33}$$

Combining Eq. (7.15) with Eq. (7.33) and substituting $(n'_x = n_x + D, n'_r = n_r)$ yields the following:

$$\delta E_{xr}(n_r) = -D(n_r + 1)\,\hbar\omega_r\,\hbar\omega_x/(4U_0). \tag{7.34}$$

Since the above expression is linear in n_r, and since the strength of carrier transitions between radial trap states is independent of the initial occupation number n_r, it is clear that $p_{xr}(\delta E_{xr})$ will simply be the occupation probability for the state $|n_r\rangle$. The result is similar to Eq. (7.20):

$$p_{xr}(\delta E_{xr}) = \frac{1}{Z_r^2 \Delta_r}\, v(\delta E_{xr})\, e^{-v(\delta E_{xr})}, \tag{7.35}$$

where $v(\delta E_{xr}) \equiv -\delta E_{xr}/[k_B T\,\hbar\omega_x\,D/(4U_0)] \geq 0$. The expression for $p_{xx}(\delta E_{xx})$ will be trickier because we must incorporate the extra weighting due to inhomogeneous driving of sideband transitions.

Probability Distribution for δE_{xx} Once again using the results from the discussion preceding Eq. (7.27), we can derive the following:

$$\langle n_x n_y n_z | V_{xx} | n_x n_y n_z \rangle = -\frac{(\hbar\omega_x)^2}{16U_0}(2n_x^2 + 2n_x + 1) \tag{7.36}$$

Proceeding as before, we then find:

$$\begin{aligned}\delta E_{xx}(n_x) &= -\left[2D(n_x + 1/2)(\hbar\omega_x)^2 + (D\hbar\omega_x)^2\right]/(8U_0) \\ &\approx -D(n_x + 1/2)(\hbar\omega_x)^2/(4U_0),\end{aligned} \tag{7.37}$$

7.2 Overview: Setting Up the Problem

where the approximation in Eq. (7.37) is to ignore the constant offset $(D\hbar\omega_x)^2$ (i.e., half the lattice-photon recoil energy), which contributes no broadening to the final lineshape because it is constant for all initial trap states.

The final probability density for δE_{xx} (i.e., the degree to which the energy difference δE_{xx} will be represented in the final lineshape) must incorporate both the occupation probability of state $|n_x\rangle$ (i.e., $p_x(n_x)$) and the weighting factor governing the strength of an axial sideband transition starting from $|n_x\rangle$ (i.e., $\Omega(n_x, D)^2$ from Eq. (7.32)). We should therefore write:

$$p_{xx}(n_x) \propto \Omega(n_x, D)^2 p_x(n_x). \tag{7.38}$$

For all n_x we'll have $p_x(n_x) = \frac{1}{Z_x} e^{-v(\delta E_{xx})}$, where $v(\delta E_{xx}) \equiv -\delta E_{xx}/[k_B T \hbar\omega_x D/(4U_0)] \geq 0$ as before.

Because the weighting factor $\Omega(n_x, D)^2$ has different forms for $D = \pm 1$, we get two slightly different expressions for the probability for red versus blue sidebands after substituting Eq. (7.37) into the expression for $p_x(n_x)$, multiplying by Eq. (7.32), and normalizing:

$$p_{xx}(\delta E_{xx}) = \begin{cases} \dfrac{v(\delta E_{xx}) + \Delta_x/2}{Z_x \Delta_x (1 + e^{-\Delta_x/2} Z_x)} e^{-v(\delta E_{xx})} & D = +1 \\[1em] \dfrac{v(\delta E_{xx}) - \Delta_x/2}{Z_x^2 \Delta_x} e^{-v(\delta E_{xx}) + \Delta_x/2} & D = -1. \end{cases} \tag{7.39}$$

Probability Distribution for δE_{SB} We can write the discrete probability $p_{\text{discrete}}(\delta E_{SB})$ analogously to Eq. (7.21):

$$p_{\text{discrete}}(\delta E_{SB}) = \sum_{\{n_x, n_r\}_{\delta E_{SB}}} p_{xx}(\delta E_{xx}) p_{xr}(\delta E_{xr}), \tag{7.40}$$

where $\sum_{\{n_x, n_r\}_{\delta E_{SB}}}$ represents a sum over all pairs of (n_x, n_r) which ensure that $\delta E_{SB}(n_x, n_r) = \delta E_{xx}(n_x) + \delta E_{xr}(n_r)$.

An analytical form for this lineshape can once again be computed by evaluating the above sum in the continuum limit, i.e.

$$\bar{p}[v(\delta E)] = \lim_{\Delta_x, \Delta_r \to 0} \frac{p_{\text{discrete}}(\delta E_{SB})}{\Delta_x \Delta_r}. \tag{7.41}$$

To evaluate this limit, we first note that when we compare Eqs. (7.35) and (7.39), we see that all three expressions simplify to the same result in the continuum limit ($\Delta_{x,r} \to 0$):

$$\bar{p}_{xi}[v(\delta E_{xi})] = \lim_{\Delta_i \to 0} \frac{p_{xi}(\delta E_{xi})}{\Delta_i} = v e^{-v} \tag{7.42}$$

for $i = x, r$ and $D = \pm 1$. The final answer is then just the following convolution:

$$\bar{p}[v(\delta E)] = \int_0^\infty \bar{p}_{xx}(v_{xx})\bar{p}_{xr}(v(\delta E) - v_{xx})dv_{xx} = \begin{cases} \frac{1}{6}v^3 e^{-v} & v \geq 0 \\ 0 & v < 0, \end{cases} \quad (7.43)$$

where $v \equiv -\delta E/[k_B T \hbar \omega_x D/(4U_0)] \geq 0$.

An appropriately scaled version of Eq. (7.43) (along the lines of the carrier lineshape from Eq. (7.30)) was used to fit the sidebands shown in Fig. 7.1b. The reason getting this lineshape right was so critical is because it significantly affects the determination of the axial trap frequency, which we operationally define as half the distance between the "start" of the sideband lineshapes on either side of the carrier. An inspection of Fig. 7.1b shows that if we were to naively use the "center of mass" of the sideband lineshape, our calculation of the axial trap frequency would be wrong by as much as 10%. Note also that for $D = \pm 1$, the "sharp edge" of the lineshape is always furthest from the carrier, since $v(\delta E) \propto -\delta E/D$.

With a functional sideband lineshape in hand, we can determine the axial trap frequency. However, in order to determine the polarizability ratio we must also measure the light shift of the carrier transition under non-magic conditions. This will be the last piece of the puzzle required for a temperature measurement.

7.2.2.3 Evaluating Light Shifts

The total shift W of a carrier transition is just the difference in expectation values for the energy between initial and final states:

$$W = \langle H' \rangle - \langle H \rangle \quad (7.44)$$

From Eq. (7.6) (i.e., the simple harmonic oscillator potential) we can rewrite $\langle H \rangle$ as:

$$\langle H \rangle = \hbar \omega_x \langle n_x + \frac{1}{2} \rangle + \hbar \omega_r \langle n_r + 1 \rangle - U_0 \quad (7.45)$$

Note that the evaluation of the expectation values above is slightly subtle. We might naively apply the equipartition theorem to get:

$$\hbar \omega_x \langle n_x + \frac{1}{2} \rangle = \hbar \omega_y \langle n_y + \frac{1}{2} \rangle = \hbar \omega_y \langle n_y + \frac{1}{2} \rangle = k_B T \text{ (equipartition)} \quad (7.46)$$

But we must be careful: this result is only exactly true in the limit of temperatures much larger than the trap spacing. We can calculate the exact values of $\langle n_x + \frac{1}{2} \rangle$ and $\langle n_r + 1 \rangle$ by again making use of Boltzmann occupation probabilities:

7.2 Overview: Setting Up the Problem

$$\hbar\omega_x\left(n_x + \frac{1}{2}\right) = \frac{1}{Z_x}\sum_{n_x=0}^{\infty}\left(n_x + \frac{1}{2}\right)e^{-\frac{\hbar\omega_x\left(n_x+\frac{1}{2}\right)}{k_BT}}$$

$$= \frac{\hbar\omega_x}{2}\coth\left[\frac{\hbar\omega_x}{2k_BT}\right] \quad (7.47)$$

$$\hbar\omega_r\left(n_y + \frac{1}{2}\right) = \hbar\omega_r\left(n_z + \frac{1}{2}\right) = \frac{1}{Z_r}\sum_{n_r=0}^{\infty}\left(n_{y,z} + \frac{1}{2}\right)e^{-\frac{\hbar\omega_r\left(n_{y,z}+\frac{1}{2}\right)}{k_BT}}$$

$$= \frac{\hbar\omega_r}{2}\coth\left[\frac{\hbar\omega_r}{2k_BT}\right]$$

Still assuming carrier transitions ($n_i = n'_i$), we can use the result above (and Eqs. (7.7) and (7.8)) to evaluate the total light shift. It will consist of three distinct parts, which we'll label W_0, W_x, and W_r:

$$W = W_0 + W_x + W_r \quad (7.48)$$

$$= \left(1 - \frac{\alpha'}{\alpha}\right)U_0 + \left(\sqrt{\frac{\alpha'}{\alpha}} - 1\right) \times \left(\frac{\hbar\omega_x}{2}\coth\left[\frac{\hbar\omega_x}{2k_BT}\right] + \hbar\omega_r\coth\left[\frac{\hbar\omega_r}{2k_BT}\right]\right)$$

Note that in the limit of large temperatures where equipartition is valid ($\hbar\omega_{x,r} \ll k_BT$) the above equation reduces to the following simplified form:

$$W \approx \left(1 - \frac{\alpha'}{\alpha}\right)U_0 + 3\left(\sqrt{\frac{\alpha'}{\alpha}} - 1\right)k_BT \quad (7.49)$$

Equations (7.48) and (7.49) imply that the total light shift can be divided into two parts: a thermal component ($W_x + W_r$) and non-thermal component (W_0). The non-thermal component U_0 can be extracted easily by fitting a spectrum with Eq. (7.28), since the "edge" of the lineshape (located at f_0) is produced by transitions of molecules at zero temperature. Therefore let's imagine a plot of (the easily observable) f_0 vs P. Let the slope of this plot be given by L_0. Then we can relate the total frequency shift to the trap depth in the following way:

$$\left(1 - \frac{\alpha'}{\alpha}\right)U_0 = hf_0 = hL_0P. \quad (7.50)$$

7.2.2.4 Putting It All Together

Next, we can rewrite Eq. (7.7) to solve for U_0 in terms of ω_x:

$$U_0 = \frac{M\lambda^2}{2}\left(\frac{\omega_x}{2\pi}\right)^2 \quad (7.51)$$

Since U_0 is proportional to P, the above equation implies that we can rewrite ω_x as:

$$\left(\frac{\omega_x}{2\pi}\right) = \kappa P^{\frac{1}{2}}, \qquad (7.52)$$

which can be measured in straightforward way. Substituting Eqs. (7.51) and (7.52) into Eq. (7.50), and then solving for (α'/α), gives the following result:

$$\frac{\alpha'}{\alpha} = 1 - \frac{2hL_0}{\lambda^2 \kappa^2 M} \qquad (7.53)$$

This is the result we were after, i.e. a method for measuring the polarizability ratio using only frequency measurements (light shifts and axial trap frequencies). Combining this with the definition for B in Eq. (7.28),

$$B = \frac{h}{k_B T (\sqrt{\alpha'/\alpha} - 1)}, \qquad (7.54)$$

allows us to rewrite the temperature T solely in terms of parameters which can be experimentally measured:

$$T_{\text{carrier}} = h \left[k_B B \left(\sqrt{1 - \frac{2hL_0}{\lambda^2 \kappa^2 M}} - 1 \right) \right]^{-1} \qquad (7.55)$$

The above equation is exact (assuming carrier transitions in a harmonic lattice with quartic corrections in the Lamb-Dicke regime), but is slightly cumbersome. But so long as $(\alpha'/\alpha) \approx 1$ (which we are assuming anyway for this derivation), we can simplify a bit:

$$T_{\text{carrier}} \approx \frac{\lambda^2 \kappa^2 M}{k B L_0} \qquad (7.56)$$

7.3 Experimental Techniques and Results

Figure 7.4 shows a comparison of temperatures determined via either the novel "carrier transition" technique described in this thesis or the old "sideband area comparison" technique often used elsewhere. The agreement between the two techniques serves as reassuring confirmation that our math is sound. The fact that the error bars are generally much smaller for the new technique serves as an advertisement for its future use.

In the following sections, I will describe the steps necessary to produce a clean, thermally-broadened lineshape suitable for publication and analysis using our (imperfect) experimental setup. I will then discuss how these clean lineshapes can be used to investigate the hidden thermal properties of our molecules.

7.3 Experimental Techniques and Results

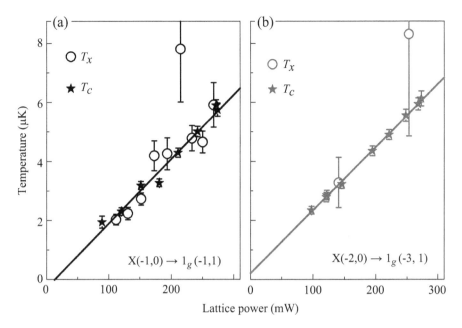

Fig. 7.4 A comparison of the temperatures of molecules as determined either by fitting the carrier lineshape in a non-magic trap (solid stars) or comparing the areas of red and blue sideband lineshapes in a magic trap (open circles). (**a**) Temperature of molecules prepared in the $X(-1,0)$ state vs lattice trapping power. (**b**) Temperature of molecules prepared in the $X(-2,0)$ state vs lattice trapping power. The large error bars on the open circles are mainly due to the difficulty in extracting the area of low-contrast sideband lineshapes. Our relative inefficiency in creating $X(-2,0)$ molecules compared to $X(-1,0)$ molecules causes this issue to be much worse in (**b**). Adapted from [8]

7.3.1 Recording High S/N Lineshapes

There are several experimental factors which make the process of obtaining clean, lattice-broadened lineshapes something of an art. Low signal-to-noise, laser frequency drift, fluctuating lattice power, and imperfectly-stabilized magnetic fields all combine to warp the recorded lineshape away from the theoretical expectation. However, some of these effects can be controlled, and others corrected for.

7.3.1.1 Stabilizing Lattice Power

Before recording any data, it is essential that the lattice power be stabilized to high precision. The reason this is so essential is that the critical quantity being measured, the "linewidth" of a thermally-broadened transition, will generally equal only a fraction of the total light shift, the exact value being determined by relationship of trap depth to molecule temperature. In our experiment, we've found that the

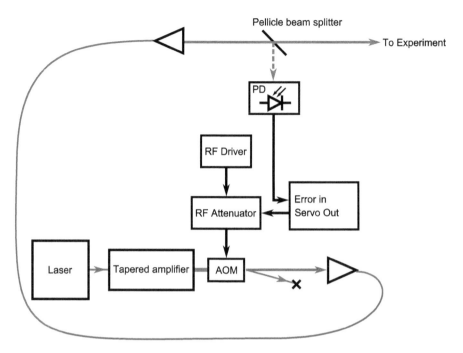

Fig. 7.5 A schematic illustrating the important components used in our lattice power stabilization scheme. Note that in order to reduce noise from interference effects, we found that we needed to sample the beam with a pellicle beam splitter

FWHM of the transition is related to the total light shift L_0 approximately according to FWHM $\approx 0.3 \times L_0$. Clearly if the lattice power drifts, it becomes impossible to record the true lineshape, since the recorded spectrum will be a superposition of several (narrow) features occurring at slightly different total light shifts.

The scheme we use to stabilize our lattice power is schematically represented in Fig. 7.5. Because we form our optical trap by retro-reflecting a laser beam, interferences are critically important to understand and control for, both in forming the potential and in sampling the lattice power. We learned by trial and error that it is necessary to use a fiber with an angled output face in order to reduce unwanted interferences due to reflections from the fiber. When the lattice power entering the chamber was viewed with an IR viewer, the use of an angled output face fiber had the effect of substantially reducing "flicker." i.e. high-frequency fluctuations in the power caused by interfering reflections.

Sampling the lattice power is also a surprisingly delicate task, since the sampler combined with a retro-reflected laser beam can form what is essentially an interferometer. We experimented with many different types of beam samplers and sampling locations in order to find a combination which produced a stable sampled beam power. The most important diagnostic for the stability of the sampled beam power was determining if the noise in the sampled power changed depending on

7.3 Experimental Techniques and Results

whether or not the retro-reflected beam was blocked. We finally settled on using a pellicle beam splitter (Thorlabs BP108) to sample the light, as indicated in Fig. 7.5. Surprisingly, our first attempt, using a D-shaped mirror to pick off a tiny fraction of the lattice power, also produced significant interference noise.

7.3.1.2 Eliminating "Cavity Drift" and "Signal Drift"

Figure 7.6a shows a spectrum depicting the loss of $X(-1, 0)$ molecules as a laser is swept across the $X(-1, 0) \rightarrow 1_g(-1, 1)$ resonance. The figure shows the superposition of seven individual traces taken one after the other. In this

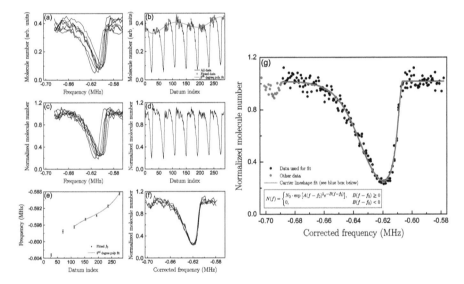

Fig. 7.6 An illustration of the process used to correct for cavity and signal drift after recording a dataset. Note that the cavity drift during this dataset of $\sim$3.3 kHz/min was atypically large, and that more realistic values for the cavity drift in 2016 are on the order of $\sim$100 Hz/min. (**a**) An uncorrected set of seven traces taken one after another, and superimposed onto the same plot. This data depicts the loss of $X(-1, 0)$ molecules as a laser is swept across the $X(-1, 0) \rightarrow 1_g(-1, 1)$ resonance. (**b**) The same data plotted against datum index (duty cycle = 0.821 ms). The red curve is a fifth-order polynomial fit to the blue points, which were chosen to be far from resonance. (**c**) The "normalized" data (i.e. the raw data divided everywhere by the value of the polynomial fit) vs frequency. (**d**) The same normalized data plotted vs datum index. (**e**) Using the carrier lineshape, each iteration was fitted in order to extract a value for f_0. This value of f_0 was then plotted vs datum index and fitted with a third-order polynomial. The first two traces were neglected for this and the following analyses because they appeared qualitatively different from the following five. (**f**) The amplitude-corrected data plotted vs "corrected frequency," defined so as to squeeze the aforementioned third-order polynomial fit into a flat line. Notice the (previously unresolvable) bump on the left-hand side of the plot, revealing a red sideband. (**g**) The fully amplitude- and frequency-corrected data plotted vs corrected frequency, and fitted with the carrier lineshape (Eq. (7.30))

example, the experimental duty cycle (i.e., the time required to record a single datum) is 821 ms, with each lineshape consisting of ∼42 data points, meaning that approximately 34 s is required to record a single lineshape. As is apparent from the figure, over the course of the ∼4 min required to record seven traces, both frequency of the spectroscopy laser and the amplitude of the signal drift in a way unrelated to the physics of the transition under investigation. These effects must be corrected for in order to extract the true lineshape.

Correcting for "Signal Drift" To determine the "shape" of the amplitude drift, we can look for a spectral feature which should have the same amplitude for each trace. A safe bet is to use simply those points which are off-resonance from either carrier or sideband transitions. These points are then plotted vs time, and a smooth function is fit to the data. We chose to use as our fitting function the lowest-order polynomial which could reasonably be said to capture all the qualitative features of the amplitude drift. This fitting process is depicted in Fig. 7.6b.

This drift is then corrected for in the data by "normalizing" (i.e. "dividing") the entire data set by the value of the polynomial at every point. The final result, plotted against frequency and time, is shown in Fig. 7.6c, d, respectively.

By inspection, it's obvious that there is something qualitatively different about the first two traces, which dip deeper and more sharply than the following five. It's possible that probe power, which is not actively stabilized, was larger for these traces before it hopped to a more stable lower value. For this reason, in the following analysis these first two traces have been removed.

Correcting for "Cavity Drift" Cavity drift is corrected using a method similar to those already described. Using the amplitude-corrected data, the full data set is divided into subsets containing a single lineshape each. The derived carrier lineshape is then fit to the data, and the fit-determined value of f_0 for each lineshape is plotted vs time. As with amplitude correction, the lowest order polynomial necessary to capture all qualitative features of the drift is used to fit the data.

The final frequency correction is then applied so as to ensure that the location of f_0 remains constant in time. This process and the results are illustrated in Fig. 7.6e, f.

The lineshape depicted in Fig. 7.6f is the final result of these machinations. Compared to the unprocessed data depicted in Fig. 7.6a, it is now much easier to identify subtle features in the spectrum, such as the small lattice cooling sideband on the left-hand side of the figure. Figure 7.6g depicts the final processed data fitted with the carrier lineshape described by Eq. (7.30). Points (in red) which are clearly part of the cooling sideband, and therefore not part of the carrier lineshape, have been removed from the fit. Note that in order to fit the data as shown in Fig. 7.6g, it is necessary to either first take the natural log of the data before fitting, or instead fit the data as shown with a function taking into consideration linear probe absorption, such as Eq. (7.30):

7.3.2 Identifying Sources of Molecular Heating

Clean lattice-broadened carrier lineshapes fitted with Eq. (7.30) allow a direct probe of the molecular temperature, and therefore offer a window into the dynamics of the molecule creation process. Figure 7.7 shows measurements of the molecule temperature under various experimental conditions. Several features deserve elaboration.

Molecule Temperature Is ∼2× Atom Temperature Figure 7.7a shows the molecule temperature vs lattice power for various durations of the photoassociation pulse. Also shown (green points) is the temperature of the atomic cloud before photoassociation, as determined by time-of-flight imaging of the atom cloud after expanding for several ms upon release from the optical lattice. Even for a photoassociation pulse as short as 40 μs, only twice as long as the lifetime of the atomic $^3P_1$ state, the molecules we produce are nearly twice the temperature of the atom cloud.

Note that our lattice is kept on during blue MOT and red MOT cooling, and that the atomic temperature is highly dependent upon the overlap of the red MOT position with the lattice beam. A consequence of this dependence is that noise in the magnetic field used for red MOT loading is imprinted onto both the number and temperature of our atom cloud. The green points in Fig. 7.7a are the average of atom cloud temperature measurements taken at the beginning and the end of the data-taking session. This was done in order to disentangle heating due to photoassociation pulse length from heating due to a drifting magnetic field.

Longer Photoassociation Pulse Times Result in Hotter Molecules This trend is apparent in Fig. 7.7a, b. Note that the duration of the photoassociation pulse is closely related to the number of molecules produced, and so the total signal at very small photoassociation pulse lengths is tiny. This is exactly the regime where carrier thermometry thrives when compared to other methods, since it is difficult to resolve red sidebands when plagued by low signal.

Off-Resonant Scattering of Lattice Photons Causes Additional Molecule Heating This trend is shown in Fig. 7.7c, where molecules produced by a 1.2 ms photoassociation pulse are held for various lengths of time before being probed. Investigations of this process will likely be important for future molecule clock experiments, as it could end up becoming a limiting factor for achieving long coherence times.

Molecules Can Be "Cooled" by Tuning a Probe to the Hot "Tail" of the Broadened Carrier Lineshape This process is demonstrated in Fig. 7.7d. In the plot shown, molecules are first pumped into the $X(-1, 0)$ state. Next, a 5 ms probe pulse is tuned somewhere in the red tail of the $X(-1, 0) \rightarrow 1_g(-1, 1, 0)$ transition

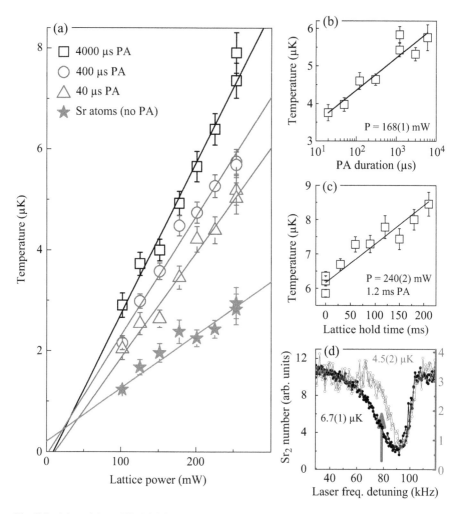

Fig. 7.7 Adapted from [8]. (**a**) Black, blue, and red points show molecule temperature vs lattice power for three different photoassociation pulse durations. Green points show the temperature of the atomic cloud, calculated as the average of measurements taken at the beginning and the end of the data-taking session. (**b**) Molecular temperature plotted vs photoassociation pulse duration. (**c**) Molecular temperature plotted vs lattice hold time, showing few-μK heating over the course of several hundred ms. (**d**) "Carrier cooling" (or perhaps more accurately, "hole-burning"), whereby hot molecules are depleted by a 5 ms probe pulse tuned to the hot tail of the lattice-broadened lineshape, resulting in a lower overall temperature for the molecules remaining in the trap

in order to deplete the hotter molecules. Finally, a second probe is pulsed, and eventually swept across resonance in order to record the lineshape and therefore the temperature.

Note that unlike in traditional resolved sideband cooling, molecules pumped out of the red tail will generally be lost rather than decay down to cooler states.

Interestingly, we also measured the molecular temperature before and after driving the *axial red sideband transition* and found no observable temperature chance of the molecules, whereas "carrier cooling" produced a very noticeable change in the lineshape.

References

1. Blatt, S., Thomsen, J., Campbell, G., Ludlow, A., Swallows, M., Martin, M., Boyd, M., Ye, J.: Rabi spectroscopy and excitation inhomogeneity in a one-dimensional optical lattice clock. Phys. Rev. A **80**(5), 052703 (2009)
2. Boyd, M.: High precision spectroscopy of strontium in an optical lattice: towards a new standard for frequency and time. PhD thesis, University of Colorado (2007)
3. Friebel, S., D'andrea, C., Walz, J., Weitz, M., Hänsch, T.: CO_2-laser optical lattice with cold rubidium atoms. Phys. Rev. A **57**(1), R20 (1998)
4. Huntemann, N., Sanner, C., Lipphardt, B., Tamm, C., Peik, E.: Single-ion atomic clock with 3×10^{-18} systematic uncertainty. Phys. Rev. Lett. **116**(6), 063001 (2016)
5. Katori, H., Takamoto, M., Pal'Chikov, V., Ovsiannikov, V.: Ultrastable optical clock with neutral atoms in an engineered light shift trap. Phys. Rev. Lett. **91**(17), 173005 (2003)
6. Lett, P., Watts, R., Westbrook, C., Phillips, W., Gould, P., Metcalf, H.: Observation of atoms laser cooled below the doppler limit. Phys. Rev. Lett. **61**, 169–172 (1988)
7. Lombardi, M., Heavner, T., Jefferts, S.: NIST primary frequency standards and the realization of the SI second. J. Meas. Sci. **2**(4), 74–89 (2007)
8. McDonald, M., McGuyer, B., Iwata, G., Zelevinsky, T.: Thermometry via light shifts in optical lattices. Phys. Rev. Lett. **114**(2), 023001 (2015)
9. McKay, D., DeMarco, B.: Cooling in strongly correlated optical lattices: prospects and challenges. Rep. Prog. Phys. **74**(5), 054401 (2011)
10. Nicholson, T., Campbell, S., Hutson, R., Marti, G., Bloom, B., McNally, R., Zhang, W., Barrett, M., Safronova, M., Strouse, G., Tew, W., Ye, J.: Systematic evaluation of an atomic clock at 2×10^{-18} total uncertainty. Nat. Commun. **6**, 6896 (2015)
11. Ost, L.: NIST launches new US time standard: NIST-F2 atomic clock (2014)
12. Ramsey, N.: A molecular beam resonance method with separated oscillating fields. Phys. Rev. **78**(6), 695 (1950)

Chapter 8
Photodissociation and Ultracold Chemistry

8.1 A ZLab History of Photodissociation Measurements

As early as 2012, we had noticed some strange features in the shapes of atom clouds formed from the photodissociation of our ultracold ^{88}Sr$_2$ molecules. With a single imaging camera oriented perpendicular to our lattice trapping axis, a cloud of trapped molecules appears as an elongated "cigar" shape. When these molecules were photodissociated, the resulting fragments would expand outward with a kinetic energy determined by the dissociating light's frequency above threshold. At the time, we weren't aware of any mechanism which might cause the photofragments to exhibit an angular dependence, particularly when the initial state was spherically symmetric with $J = 0$. And yet *something* strange was happening—the "fuzz" of the dissociated atom cloud seemed to change shape with changing dissociating laser frequency. Figure 8.1 shows an early example of what we were seeing.

As is evident in the figure, clouds which at small frequencies appeared to be concentrated above and below the lattice axis at small dissociation frequencies were soon replaced with a "fuzz" filling in the middle at larger frequencies. We could think of several possible mechanisms which might cause this fuzz. Our best guess was that there might exist some $J = 1$ quasibound state (or shape resonance) $\sim$30 MHz above threshold to which the dissociation laser was inadvertently transferring population, and which was creating low-energy fragments when it spontaneously decayed. To investigate this possibility, we turned to our theory collaborators Robert Moszynski and Wojtek Skomorowski, and asked them to use their ab initio model to try to predict where these purported shape resonances might occur. But when no matter how they tweaked their model no shape resonances emerged. So we were stuck. Our best guess for the cause of the "fuzziness" seemed to be unsupported by theory, and we didn't yet know of a way to check our next best guess that the photofragment angular distribution might be anisotropic and depend upon laser frequency.

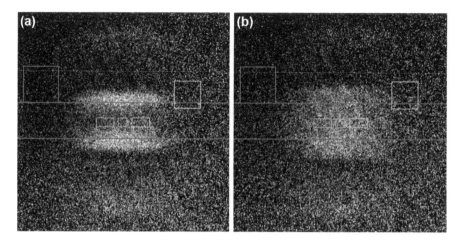

Fig. 8.1 Images from March 20, 2013 showing photodissociation of $X(v = -2; J = 0, 2)$ molecules at (**a**) 20 MHz and (**b**) 30 MHz above threshold. The noticeable "filling" of the middle of the picture was at first assumed to be due to the population of a quickly-decaying $J = 1$ shape resonance, but was later attributed to a changing photofragment angular distribution

So the situation stood until mid-2014, when Master's student Florian Apfelbeck joined our lab from the Ludwig Maximilian University of Munich. After having spent several months working on laser construction for a different experiment in our group, Florian was eager for a chance to do experiments which might yield more data which could eventually be written up in a Master's thesis. I recognized that this offered a perfect opportunity to steal away a smart, motivated student to help solve a problem Bart McGuyer and I had struggled with for several months, and so I suggested that Florian might be interested in the following project.

The primary impediment to extracting angular information from our photodissociation clouds was the fact that the atoms emerged not from a single point, but from an elongated cigar-like shape. This caused the cloud to become blurred out horizontally, hiding angular information. However, if we were to install another camera to view the cloud from along the lattice axis, then the initial distribution of molecules would appear as a point. Then the locations of the photofragments on the CCD would depend only upon the angle at which they emerged, and not on their initial location within the cloud. Figure 8.2 shows examples of both the initial molecule distribution and a photofragment cloud from these two perspectives.

After several months of hard work and failed attempts, we were finally able to integrate another camera system into our experiment, using D-shaped mirrors to bounce an absorption imaging beam as close as possible to the lattice axis and a microscope objective with a working distance just long enough to accommodate our $\sim$22 cm diameter vacuum chamber. It took several months more to develop techniques for maximizing the signal-to-noise of our experimental images. Figure 8.3 shows a timeline of representative images demonstrating improvements in our imaging ability.

8.1 A ZLab History of Photodissociation Measurements

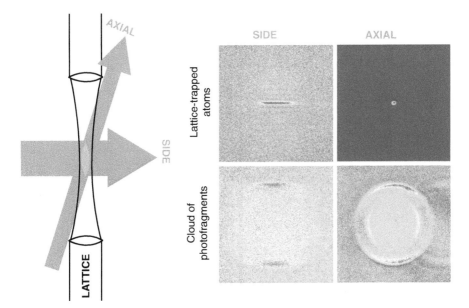

Fig. 8.2 The orientation of the imaging beam (illustrated schematically at left) dramatically affects the appearance of the cloud of photofragments. Here are shown photofragments produced from a mixture of $X(-1, 0)$ and $X(-1, 2)$ molecules, imaged from either the side or nearly along the lattice axis. The energy of the photofragments is the same in both pictures

With this new imaging ability, we've found that we can access an entirely new spectrum of questions to ask about the properties of molecules. This can be attributed to the fact that the data we record in photodissociation studies are *qualitatively different* than the kind of data so far described in previous chapters. Up until now, we've described a multitude of properties of molecular states which can be measured in order to build up a coherent understanding of a molecule and its environment, including binding energies, Zeeman shifts, transition strengths, lineshapes, and state lifetimes. But in order to measure any one of these quantities, the same basic technique is applied: a laser is swept across a transition, and the number of atoms remaining after some manipulation is measured. Therefore every one of these quantities can be traced back to a measurement of *atom number*. In photodissociation, however, we record a fundamentally different kind of information in the *direction* of photofragments upon dissociation. And just as the advent of a new kind of telescope often heralds unforeseen discoveries in astronomy, so has our ability to record photofragment direction led to surprising discoveries in the physics of ultracold molecules.

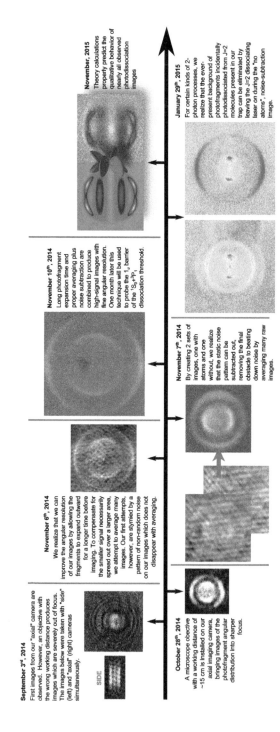

Fig. 8.3 A timeline outlining improvements made in our axial imaging scheme, leading eventually to the ability to precisely extract photofragment angular information and compare with theory calculations

8.2 A Slightly Broader History of Photodissociation Measurements

8.2.1 Early (Classical) Theory

Photodissociation is a tool which has been used by chemists since the 1960s in order to reveal information about a molecule's structure. The realization that such measurements could be useful came in 1963, when Richard Zare and Dudley Herschbach discovered [16] that the angular distribution of photofragments produced in a photodissociation reaction could be described by the expression

$$I(\theta) \propto 1 + \beta_2 P_2(\cos\theta), \beta_2 \in (-1, 2), \tag{8.1}$$

where θ is the angle of the emitted photofragments with respect to the dissociating light's polarization axis and $P_2(\cos\theta)$ is the second Legendre polynomial. If the dissociating light is linearly polarized and the axial recoil approximation is assumed (i.e., fragments are assumed to recoil along the direction of the internuclear axis upon dissociation), then the value of β in the above equation can be directly related to the orientation of the molecule's transition moment, with $\beta = 2$ corresponding to *parallel* transitions (e.g., in diatomic molecules those for which the transition moment is parallel to the internuclear axis) and $\beta = -1$ corresponding to *perpendicular* transitions [14].

8.2.2 The First Experiments

At the time Zare and Herschbach made their prediction, anisotropic photofragment angular distributions had not yet been experimentally observed, much less quantitatively analyzed. This would have to wait until 4 years later, when Jack Solomon (interestingly enough, *also* a graduate student at Columbia) made the first experimental observation of an anisotropic photofragment angular distribution using a method called "photolysis mapping" [13]. This early experiment was performed by coating a glass hemisphere with tellurium and surrounding the hemisphere with a gas of molecular iodine (I_2). When a photodissociation laser was passed through the hemisphere, iodine molecules would dissociate into atoms, hitting the hemisphere anisotropically in a way reflecting the character (parallel or perpendicular) of the photodissociative transition. Because the tellurium coating would react strongly with iodine atoms, but not at all with iodine molecules, the anisotropy could be measured by examining the thickness of the tellurium coating at different points on the hemisphere.

Needless to say, this was a painstaking way of measuring anisotropic photofragment angular distributions. Later experiments would make use of molecular beams, where a collimated beam is subjected to a polarized dissociation laser and the fragments are collected on a plate some distance away. The first experiments of this kind were performed by Kent Wilson and company in the late 1960s [6], and experiments of this same basic architecture are still carried out today. For an excellent summary of these early results, see Zare's 1972 review [14].

8.2.3 Difficulties in Comparing Experiment to a Fully Quantum Mechanical Theory

Almost immediately after Zare and Herschbach's 1963 result, theorists began refining photofragment angular distribution (PAD) predictions for different systems using the full machinery of molecular quantum mechanics. However, comparison of theory with experiment was difficult. This is because as systems become complex, full quantum mechanical calculations become extremely difficult, and the hot molecular beams serving as the starting points for most photodissociation experiments were certainly complex. High temperatures imply a multitude of quantum states being represented in the initial sample, and a fully quantum mechanical calculation would have to sum over matrix elements connecting every represented initial state to every allowed final state, with weighting factors which might only be determinable empirically.

In an effort to simplify matters, it was suggested in the late 1980s [7, 15] that rather than evaluating a complicated expression summing over a multitude of channels, the photofragment angular distribution could be summarized by a *quasiclassical formula* of the form

$$I(\theta) = \frac{\sigma}{4\pi} P_{JKM}(\cos\theta)[1 + \beta_2 P_2(\cos\theta)], \tag{8.2}$$

where $P_{JKM}(\cos\theta)$ is a function representing the "shape" of the initial state and $[1 + \beta_2 P_2(\cos\theta)]$ is the usual probability of photodissociation occurring for parallel or perpendicular transitions.

The above formula has intuitive appeal. It *makes sense* that the direction of recoiling fragments should be influenced by the initial orientation of the molecule. However, at the time this result was published, it was controversial, since it appears to have a dramatically different form than the fully quantum mechanical solution.

Several experiments have been performed which appear to show qualitative agreement with the quasiclassical formula [2, 3]. And after a few more years of theory investigations, it became clear why this might be. Perhaps the first detailed investigation of the applicability of the quasiclassical formula was performed by Tamar Seideman in 1996 [12]. (Another excellent analysis was performed by Beswick and Zare in 2008 [4].) Seideman's conclusion was that the quasiclassical and fully quantum mechanical calculations reduce to the same result only if it is assumed that "the scattering wavefunctions are independent of the rotational branch and that the transition dipole vector is very simple." The first assumption in this statement is self-explanatory, while the second is equivalent to the requirement that the angular momentum projection Ω along the internculear axis for the initial and final states is a good quantum number [4].

Is it possible to perform experiments in a regime where the quasiclassical approximation would be expected to break down? And even if the answer were yes, can experimental precision be made high enough to unambiguously detect it? Up until the middle of 2016 the answer had been a resounding..."maybe." While several early experiments showed consistency with the quasiclassical interpretation, several others hinted at violations. A result from 1997 by Pipes et al. examining the Doppler profiles of photodissociated fragments implied that the quasiclassical approximation should fail when interference effects become important (i.e., when the system cannot be described by an incoherent sum over states) [10]. An unambiguous exploration of the breakdown of the quasiclassical approximation would require experiments to be performed in a regime where quantum interference effects dominate the reaction mechanism, and where those interference effects could be cleanly observed. Our experiment provides the first glimpse of photochemical reactions in this exciting regime.

8.3 Photodissociation of Ultracold ^{88}Sr$_2$ Molecules in an Optical Lattice

To study photodissociation reactions in a fully quantum mechanical way, we start with ultracold molecules whose initial internal state can be completely defined. As was discussed in Chaps. 3 and 4, vibrational and rotational numbers can be selected by tuning the frequencies of our probe lasers, while magnetic sublevel can be selected by splitting rotational levels with a magnetic field. The fragments are emitted with a velocity depending on the energy of the dissociation laser above threshold, and with angles θ and ϕ defined with respect to the molecular quantization axis (Fig. 8.4a). The fragments expand outward in a spherical shell, which is projected onto a 2D plane via absorption imaging (Fig. 8.4b). In order to most clearly resolve the angular direction of the photofragments, the absorption imaging beam is aligned as closely as possible with the lattice axis (Fig. 8.4c and discussion in Sect. 8.1).

8.3.1 Imaging Photofragment Angular Distributions (PADs)

In order to achieve high sensitivity to photofragment direction (unclouded by blurring from the initial molecule distribution), we allow the fragments to expand for a distance much larger than the initial trap size and ensure that the photodissociation

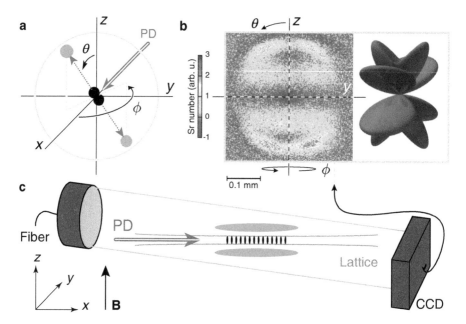

Fig. 8.4 Photodissociation of diatomic molecules in an optical lattice. (**a**) A homonuclear molecule (black circles) producing fragments (green circles) with well-controlled speeds forms a Newton sphere. The distribution of the fragments on the sphere surface is parameterized by a polar angle θ relative to the z-axis and an azimuthal angle ϕ relative to the x-axis in the xy-plane. The photodissociating (PD) light propagates along $+x$. (**b**) An experimental image of the fragments corresponds to the Newton sphere projected onto the yz-plane. This particular image is one of many we observed that is highly quantum mechanical in nature, manifest in the distinct lack of cylindrical symmetry about the z-axis. (**c**) The fragments (green ovals) are detected by absorption imaging using a charge-coupled device (CCD) camera and a wide light beam from an optical fiber. The photodissociating light is co-aligned with the lattice axis along x. The imaging light is nearly coaligned with x (a small tilt is present for technical reasons). A magnetic field can be applied along the z-axis. Adapted from [8]

pulse duration is much smaller than the expansion time. However, both of these requirements must be balanced with the requirement for high signal to noise. Large expansion times dilute small signals across a large number of CCD pixels, while short-duration photodissociation pulses limit the number of molecules which ultimately dissociate. Because of this, it is imperative to remove as many sources of imaging noise as possible. For an excellent discussion of how we achieved this in our experiment, see Florian Apfelbeck's Master's Thesis [1]. For convenience, Fig. 8.5 (reproduced with permission from the author) outlines the process we used to achieve high signal-to-noise from averages of hundreds of experimental shots.

8.3.2 Extracting Quantitative Angular Information

The general form for the photofragment angular distribution (PAD) produced by a photodissociation reaction can be parameterized by the following formula [11]:

$$I(\theta,\phi) \propto \sum_{L=0}^{L_{\max}} \sum_{M=-L}^{+L} \beta_{LM} Y_{LM}(\theta,\phi), \tag{8.3}$$

where $L_{\max}$ is twice the angular momentum of the highest L output channel and the β_{LM} coefficients can be complex functions of ϕ (but will be scalar constants in situations where the PAD is cylindrically symmetric). Comparing the results of a photodissociation experiment with theory can then be achieved by comparing the measured versus calculated β_{LM} parameters, which requires being able to extract information about a 3D angular distribution (parameterized by angles θ and ϕ) from a 2D image of fragment positions (parameterized by x and y).

If the PAD possesses cylindrical symmetry about the z-axis, then we can achieve this through *Abel inversion*, specifically implemented through a software implementation of the pBasex algorithm by Luka Pravica. Figure 8.6 (reproduced with permission from Florian Apfelbeck) outlines the process we use to extract angular information from 2D images.

Figures 8.7 and 8.8 demonstrate two examples where we plot β-coefficients extracted from experimental images and compare with theory calculations. However, the Abel inversion algorithm is only applicable to situations for which cylindrical symmetry exists about an axis lying in the imaging plane. This condition holds when the polarization vector is parallel to the quantization axis, but generally fails for the equally interesting case of polarization perpendicular to the quantization axis.

In order to compare theory predictions to this special case, we use theory-provided β-coefficients to produce simulated images, and then compare these simulated images to experimental results. Figure 8.9 shows a library of such comparisons for both parallel and perpendicular polarization directions, demonstrating excellent theory agreement in every case.

8.3.3 Kinematics of a Photodissociation Reaction

The molecules we dissociate do not exist in free space, but rather are trapped in a lattice with a total depth of ~ 1 MHz. It's fair to ask whether this confining potential would affect to the natural trajectories of the photofragments we observe. We certainly do not expect the confining potential to influence photofragment trajectories when the dissociation energy is much larger than the trap energy, since in this extreme case the potential serves only as a "bump on the road" for fragments

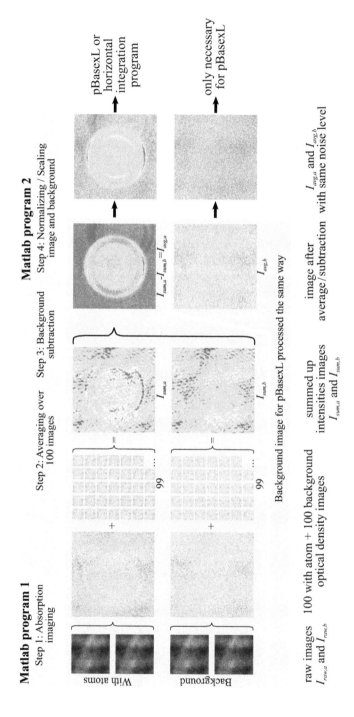

Fig. 8.5 Please view sideways. Process to increase signal-to-noise ratio. We start with raw images (left) and produce averaged and background-subtracted images (right). Reproduced with permission from Florian Apfelbeck from [1]

8.3 Photodissociation of Ultracold ^{88}Sr$_2$ Molecules in an Optical Lattice

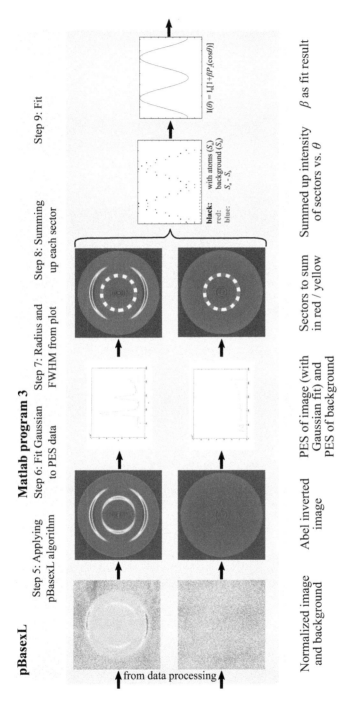

Fig. 8.6 Please view sideways. Process to retrieve β with pBasex inversion method. Reproduced with permission from Florian Apfelbeck from [1]

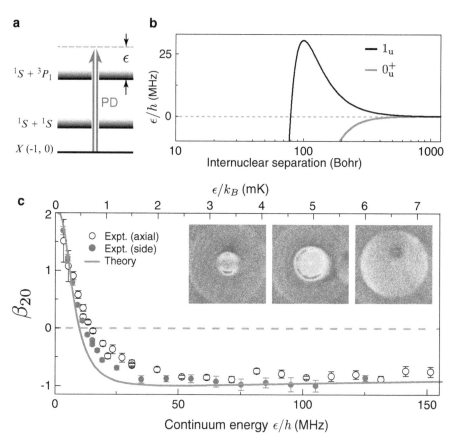

Fig. 8.7 Photodissociation to a multichannel continuum. (**a**) Schematic for PD of ^{88}Sr$_2$ in the initial ground state $X(v_i, J_i)$ to an excited continuum energy ϵ, which is subsequently expressed in MHz or in mK (via the Boltzmann constant kB). (**b**) Potential energy structure ($\lesssim 1$ mK) of the $^1S+^3P_1$ continuum, showing both of the electronic potentials (0_u^+ and 1_u) that couple to the ground state via E1 transitions [5]. (**c**) The angular anisotropy parameter β_{20} for this process measured by two imaging methods (using axial-view and side-view CCD cameras) and calculated using a quantum chemistry model. The inset images show fragments at three different energies ϵ/h labeled in MHz. The images and curves indicate a steep change in the angular anisotropy in the 0–2 mK continuum energy range. The experimental errors for axial imaging were estimated by varying the choice of center point for the pBasex algorithm and averaging the results, and for side imaging from least-squares fitting to Eq. (2) in our Nature publication [8] convolved with a blurring function to account for experimental imperfections. Reproduced from [8]

racing out of the trap. And empirically, we observe agreement in the calculated photofragment angle whether observed axially or from the side (Fig. 8.7), evidence that the distributions we observe truly are cylindrically symmetric.

Figure 8.10 plots the radius of photofragment rings from both $J = 0$ and $J = 2$ molecules observed in the data set described in Fig. 8.7. The clear square root

8.3 Photodissociation of Ultracold ^{88}Sr$_2$ Molecules in an Optical Lattice

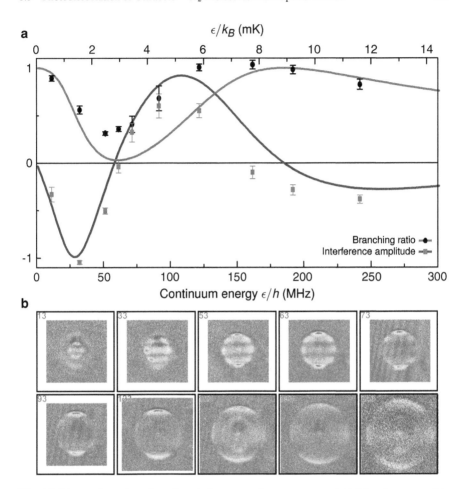

Fig. 8.8 Energy-dependent photodissociation near a shape resonance. (**a**) Molecules prepared in the $0_u^+(v_i = -3, J_i = 3, M_i = 0)$ state are photodissociated at the ground continuum. For $p = 0$, selection rules lead to a single $M = 0$ but a mixture of $J = 2, 4$. The branching ratio and interference amplitude of this mixture, as described in the text, evolve with energy and reveal a $J = 4$ (g-wave) shape resonance at ∼3 mK. The experimental data were analyzed with pBasex and errors were estimated by varying the effective saturation intensity, used to process the absorption images, within its uncertainty. The theoretical curves were calculated with a quantum chemistry model. (**b**) Images of fragments labeled by their continuum energies ϵ/h in MHz that show the evolution with energy. The faint anisotropic, energy-independent pattern with roughly the same radius as the 62 MHz image is from spontaneous decay into the shape resonance. Adapted from [8]

dependence of ring radius vs dissociation energy agrees with what is expected from energy conservation, and in Fig. 8.10d evidence of a ∼1 MHz potential barrier is clearly visible.

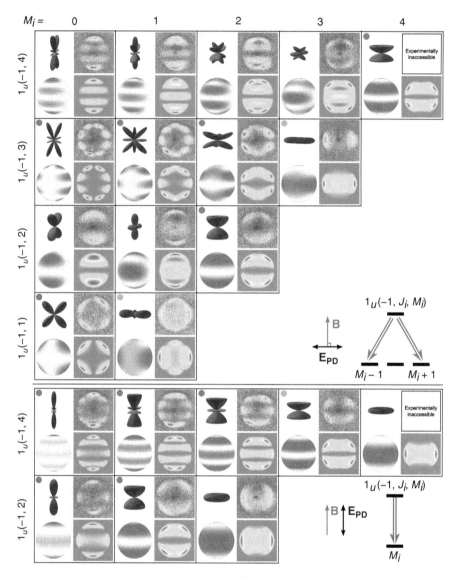

Fig. 8.9 Photodissociation of singly-excited (^{1}S+^{3}P$_1$) molecules to the ground-state continuum with energies of several millikelvin. Each row and column corresponds to molecules prepared in the indicated $1_u(v_i, J_i)$ state and M_i sublevel. ($M_i = 4$ was not accessible experimentally.) The upper and lower sections correspond to PD light polarizations $|p| = 1$ and 0, respectively, where the PD lasers electric field is E_{PD}. Within each square panel, the experimental image is on the top right, with a comparable simulation of a projected Newton sphere on the bottom right. The full sphere rendition is on the bottom left and the top left shows the mapping of the fragment detection probability at each angle onto the radial coordinate of a surface. For $|p| = 1$, matterwave interference occurs if two values of M are produced, leading to strongly ϕ-dependent patterns. For each case, the degree of agreement with the quasiclassical approximation is indicated by a colored dot, as explained in our Nature publication. Reproduced from [8]

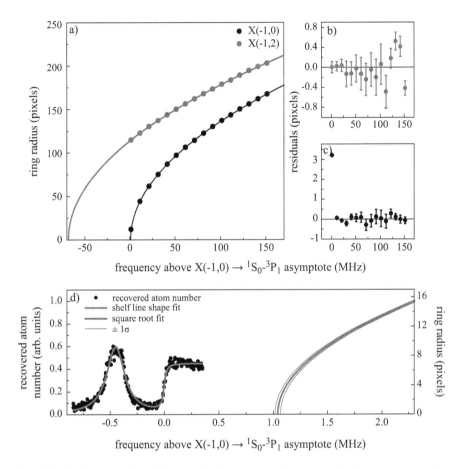

Fig. 8.10 In (**a**), the radius of the two signal rings is plotted against the frequency above the $^1S_0$-$^3P_1$ asymptote when dissociating from $X(-1,0)$. (**b**) and (**c**) show the residuals, while in (**d**) the shelf line on the left side and a zoomed in version of the $X(-1,0)$ plot in (**a**) is illustrated. The effect of the depth of the lattice becomes visible, as the dissociation fragments can only be seen after 1.04(2) MHz and not right above at the dissociation threshold. Reproduced with permission from Florian Apfelbeck from [1]

8.3.4 A Fully Quantum Mechanical Understanding of Photodissociation

In Fig. 8.9 and Extended Data Figs. 2 and 3 of our Nature paper [8] we observe many cases in which the quasiclassical approximation fails dramatically. But perhaps even more interesting is when it succeeds. It turns out that *when selection rules allow for only a single rotational state (i.e., partial wave) to be present in the output channel, the fully quantum mechanical and quasiclassical calculations yield identical predictions*. This is in keeping with Seideman's requirement that the scattering wavefunctions be "independent of the rotational branch," as well as with the conjecture by Pipes et al. that the quasiclassical approximation might fail when interference effects become important (since if there is only a single rotational state possible, no interference is possible).

Our ability to precisely populate single quantum states and manipulate them for extended periods of time is what enables these surgical, targeted studies of when different theoretical regimes should be applicable. And the relative simplicity of our ultracold, homonuclear, diatomic molecular system has enabled our theory collaborators to calculate exactly the fully quantum mechanical predictions from a quantum chemistry model, allowing us to understand from first principles the results of almost every experiment we can think to perform.

Almost... but not completely. In the next section I'll describe several sets of experimental results which (as of the writing of this thesis) continue to beguile us. I hope that these mysteries might serve as inspiration to the next generation of ultracold photodissociation researchers to jump in and develop an even deeper understanding of this simple chemical reaction.

8.4 Preliminary Data and Unresolved Mysteries

8.4.1 *Magnetic Field Dependence of Dissociation Above the $1_u/0_u$ Barrier*

One of the salient points we attempted to make in our recent Nature paper [8] was to show that in the ultracold, fully quantum state selected regime, the quasiclassical model describing the photodissociation of diatomic molecules is simply *wrong*. And the reason that model fails is because it makes the explicit assumption that the direction of outgoing photodissociation fragments depends upon the *initial orientation* of the molecule being photodissociated. While this idea has intuitive appeal (if a molecule starts off pointing in a certain direction, shouldn't the fragments continue along that direction once the bond is cut?), it hides the true nature of the physics which determines photofragments' trajectories. As we explained in our paper, what actually matters is the *superposition of rotational wavefunctions allowed in the final output channel*. And while it's true that the initial state will influence which channels are allowed in the final state through selection rules, knowledge of the initial state's "shape" is only helpful in determining the final photofragment angular distribution in scenarios for which interference effects are unimportant or where the final output channel is independent of rotational angular momentum [4].

This insight forms the nucleus of our understanding about what types of photofragment angular distributions are possible for a given experiment. It was what allowed us to explore the $J = 4$ shape resonance above the $^1S_0+{}^1S_0$ threshold

8.4 Preliminary Data and Unresolved Mysteries

by fitting photofragment angular distributions with only two free parameters (an "amplitude" and an "interference" coefficient describing the interplay between $J = 2$ and $J = 4$ spherical harmonics). It also explains why the dissociation of states with different values of Ω (0 or 1) can produce nearly identical photofragment angular distributions, despite the "shapes" of the initial states (as defined by Wigner D-functions) being dramatically different. Having developed an intuition which is so powerful in categorizing the behaviors of different photodissociation experiments, we would be very surprised to discover a case in which this intuition failed.

The behavior of one-photon dissociation of $X(-1,0)$ molecules above the $1_u/0_u$ threshold in the presence of a magnetic field, however, currently appears to be just such a case. Figure 2 from our Nature paper shows an analysis of the photofragment angular distribution at an applied magnetic field of ~ 0 G. Because the initial state is spherically symmetric, this experiment can be interpreted *semi-classically*, such that when photofragments are emitted mostly vertically we are observing a *parallel transition*, and when they are emitted mostly horizontally we are observing a *perpendicular transition*, where "parallel" or "perpendicular" refers to the orientation of the transition moment with respect to the internuclear axis. According to Zare and Herschbach's 1963 derivation, the photofragment angular distributions in these cases will be proportional to either $\cos^2\theta$ or $\sin^2\theta$ respectively, where θ is the angle between the dissociating light's polarization axis and the photofragment's emission direction. However, if our fully quantum mechanical intuition is justified, we should be able to think of this experiment in terms of shapes of the allowed *final state* rotational wavefunctions as well. And in this case, the quantum mechanical picture works fine. Linear laser polarization ensures that $\Delta m = 0$. Selection rules require $|\Delta J| = 0, 1$. Since $J = 0$ is forbidden for the 1_u and 0_u potentials, there are only two allowed channels in the final state. And as luck would have it, the shape of the $m = 0$ component of a $J = 1$, 0_u wavefunction is proportional to $\cos^2\theta$, while the shape of its counterpart 1_u wavefunction is proportional to $\sin^2\theta$. The quantum mechanical and semiclassical pictures of photodissociation in this case mesh nicely with one another.

We would *not* expect the situation to change dramatically in the presence of an applied magnetic field. For linearly polarized light parallel to the quantization axis, the selection rule $\Delta m = 0$ should still be enforced. The shapes of the wavefunctions describing the final state should *probably* not change by much. And yet when we perform this experiment, we see a discreet change in the appearance of the photofragment angular distribution. Figure 8.11 summarizes the observed behavior of this process at a variety of magnetic fields and detunings above threshold.

There are two very difficult-to-explain features in the data shown in Fig. 8.11:

1. **Photofragment angular distributions appear to be described by functions more complicated than $\cos^2\theta$.**
   As we discuss in our Nature publication (see Methods), the "shape" of the rotational wavefunction for a diatomic molecule is given by the Wigner D-function $D^J_{M\Omega}(\phi, \theta, \chi)$. For $J = 1$, it turns out that the D-functions have the simple form $\sin\theta$ or $\cos\theta$ for $\Omega = 1, 0$. Since there are only two channels possible in the final

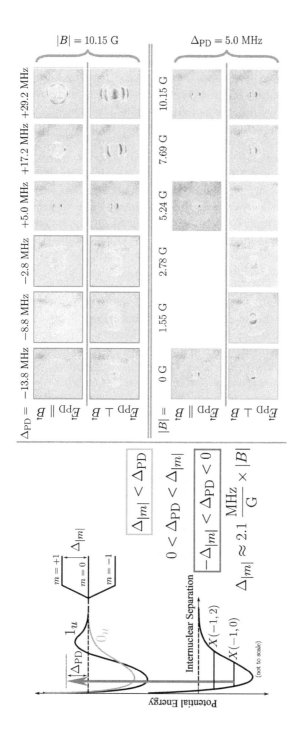

Fig. 8.11 A series of images (displayed at right) depicting the photofragment angular distributions resulting from photodissociation of $X(-1, 0)$ molecules (scheme schematically shown at left) at various magnetic fields and detunings above threshold. The colored outlines of each image indicate which magnetic sublevels are allowed, as indicated by the boxed expressions at left. In all experimental images, the faint outermost ring is the result of incidental photodissociation of $J = 2$ molecules and should be ignored. (The $J = 2$ population was imperfectly pumped out of the trap before dissociation. For details of the process, see [1].)

8.4 Preliminary Data and Unresolved Mysteries

output channel for this process, we would expect that the photofragment angular distribution could be written as the generic sum $A \cdot \cos^2\theta + B \cdot \sin^2\theta$. The pattern described by such a sum smoothly varies between two lobes of high intensity at either top and bottom or left and right, to spherically symmetric.

Indeed, at zero magnetic field this is what we see. However, once the magnetic field becomes large enough to resolve the $m = \pm 1$ sublevels, the pattern clearly becomes more complicated. Perhaps the most striking example of this is in the top-rightmost image in Fig. 8.11, showing clear evidence of *six lobes* in the photofragment angular distribution. Currently, our only idea for explaining this would be magnetic field-induced *J*-mixing, the same phenomenon responsible for enabling the control of forbidden transitions described in Chap. 5. However, this would seem to imply that the photofragment angular distribution should become more distorted as the magnetic field increases, a fact which seems to be contradicted by the lower half of Fig. 8.11.

2. **Δm selection rules in some cases appear to be broken.**
   When the dissociation laser polarization is *parallel* to the applied magnetic field direction, we would expect that for E1 transitions, $\Delta m = 0$. Conversely, when the polarization is *perpendicular* to the applied magnetic field, we should see $\Delta m = \pm 1$.

   We can identify a ring's magnetic sublevel designation by its linear Zeeman shift. This is illustrated in the expression $\Delta_{|m|} \approx 2.1 \frac{\text{MHz}}{\text{G}}$ given at the left side of Fig. 8.11. And in nearly every image where both $m = 0$ and $m = -1$ are energy-allowed (indicated by the colored borders on the images), we see two rings. The strong $m = 0$ ring when $\vec{E}_{\text{PD}} \perp \vec{B}$ might be attributed to the possibility that the $m = 0$ component of the 0_u shelf has zero linear Zeeman shift (though this too would be surprising, since it contradicts the intuition we've built up concerning 0_u bound states). The fainter $|m| = 1$ rings observed when $\vec{E}_{\text{PD}} \parallel \vec{B}$, is harder to explain.

Solving these mysteries would be extremely satisfying, because it would help confirm that we fully understand this chemical reaction in the quantum regime. But there might be another added benefit as well. Since ring radius is proportional to the square root of kinetic energy, image sets like the lower row of Fig. 8.11 can be used to perform spectroscopy on the $m = -1$ component's energy versus applied magnetic field. This would tell us the *Zeeman shift* of one of the components of the dissociation threshold, which would be a very interesting number to have, since we've already shown that the quadratic Zeeman shifts of rovibrational levels get larger and larger as they approach the dissociation threshold. It would also be interesting to calculate mixing angles for the 1_u and 0_u components of the threshold (if they could indeed be separately resolved) by measuring their linear Zeeman shifts and following the procedure we used in 2013 [9].

8.4.2 Frequency Dependence of One-Photon Dissociation of $J = 1$ States Down to $^1S_0+^1S_0$ Threshold in the Absence of Shape Resonances

As was shown in Fig. 5 of our Nature publication [8], the photofragment angular distribution produced by dissociation of the $0_u(-3, 3)$ state with linearly polarized laser light tuned just above the $^1S_0+^1S_0$ threshold will change dramatically with laser frequency. Because there are only two allowed output channels, we found that the full angular distribution can be summarized with only two free parameters R and δ in the following way:

$$|f(\theta,\phi)|^2 = |\sqrt{R}\, Y_{20}(\theta,\phi) + e^{i\delta}\sqrt{1-R}\, Y_{40}(\theta,\phi)|^2, \qquad (8.4)$$

where ϕ is the azimuthal angle and Y_{J0} is a spherical harmonic.

A plot of the value of R vs laser frequency revealed a dramatic dip at $\sim$66 MHz above threshold, which we interpreted as evidence of the presence of a $J = 4$ shape resonance. However, even when no shape resonances are thought to exist for the quantum numbers allowed in the final output channel, we *still* observe dramatic variation among photofragment angular distributions, depending upon both frequency above threshold and the initial state.

Figure 8.12 shows a collection of photofragment angular distributions recorded at four energies above threshold and starting from six different initial states. What's interesting in this figure is that photofragment angular distributions produced from nearly all initial states show a dependence upon frequency above threshold. Photodissociation of the $0_u(-3, 1)$ state is also interesting, because its obvious near-spherical symmetry implies that the $J = 2$ channel contributes only very weakly to the final output state.

Preliminary calculations by our theorist collaborators at the University of Warsaw have already begun to show qualitative agreement with our data. But careful comparison has not been made, and it will be exciting to see whether in the future more general rules can be discovered to predict the shapes of these patterns.

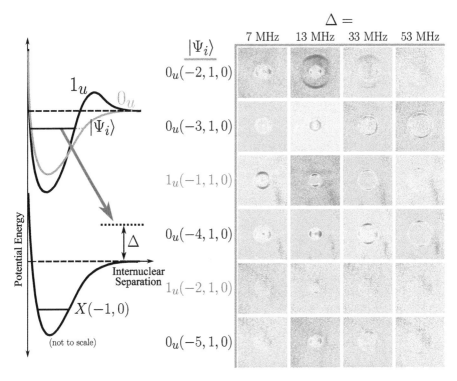

Fig. 8.12 The photofragment angular distribution produced by photodissociation of $J = 1$ excited states by a laser tuned above the $^1S_0 + {}^1S_0$ dissociation threshold is not only frequency-dependent, but dramatically different for nearly every level observed. No shape $J = 0$ or $J = 2$ shape resonances are known to exist for ^{88}Sr. Is this variation among different initial states an incidental fluke which can be explained by exact calculations of the transition moment, or is a deeper explanation possible?

References

1. Apfelbeck, F.: Photodissociation dynamics of ultracold strontium dimers. Master's thesis, Ludwig Maximilian University of Munich, Munich (2015)
2. Baugh, D., Kim, D., Cho, V., Pipes, L., Petteway, J., Fuglesang, C.: Production of a pure, single ro-vibrational quantum-state molecular beam. Chem. Phys. Lett. **219**(3), 207–213 (1994)
3. Bazalgette, G., White, R., Trénec, G., Audouard, E., Büchner, M., Vigué, J.: Photodissociation of ICl molecule oriented by an intense electric field: experiment and theoretical analysis. J. Phys. Chem. A **102**(7), 1098–1105 (1998)
4. Beswick, J., Zare, R.: On the quantum and quasiclassical angular distributions of photofragments. J. Chem. Phys. **129**(16), 164315 (2008)
5. Borkowski, M., Morzyński, P., Ciuryło, R., Julienne, P., Yan, M., DeSalvo, B., Killian, T.: Mass scaling and nonadiabatic effects in photoassociation spectroscopy of ultracold strontium atoms. Phys. Rev. A **90**(3), 032713 (2014)

6. Busch, G., Mahoney, R., Morse, R., Wilson, K.: Photodissociation recoil spectra of IBr and I_2. J. Chem. Phys. **51**(2), 837–838 (1969)
7. Choi, S., Bernstein, R.: Theory of oriented symmetric-top molecule beams: precession, degree of orientation, and photofragmentation of rotationally state-selected molecules. J. Chem. Phys. **85**(1), 150–161 (1986)
8. McDonald, M., McGuyer, B., Apfelbeck, F., Lee, C.-H., Majewska, I., Moszynski, R., Zelevinsky, T.: Photodissociation of ultracold diatomic strontium molecules with quantum state control. Nature **534**(7610), 122–126 (2016)
9. McGuyer, B., Osborn, C., McDonald, M., Reinaudi, G., Skomorowski, W., Moszynski, R., Zelevinsky, T.: Nonadiabatic effects in ultracold molecules via anomalous linear and quadratic Zeeman shifts. Phys. Rev. Lett. **111**(24), 243003 (2013)
10. Pipes, L., Brandstater, N., Fuglesang, C., Baugh, D.: Photofragmentation of M-state polarized molecules: comparison of quantum and semiclassical treatments. J. Phys. Chem. A **101**(41), 7600–7604 (1997)
11. Reid, K.L.: Photoelectron angular distributions. Annu. Rev. Phys. Chem. **54**(1), 397–424 (2003)
12. Seideman, T.: The analysis of magnetic-state-selected angular distributions: a quantum mechanical form and an asymptotic approximation. Chem. Phys. Lett. **253**(3), 279–285 (1996)
13. Solomon, J.: Photodissociation as studied by photolysis mapping. J. Chem. Phys. **47**(3), 889–895 (1967)
14. Zare, R.: Photoejection dynamics. Mol. Photochem. **4**, 1–37 (1972)
15. Zare, R.: Photofragment angular distributions from oriented symmetric-top precursor molecules. Chem. Phys. Lett. **156**(1), 1–6 (1989)
16. Zare, R., Herschbach, D.: Doppler line shape of atomic fluorescence excited by molecular photodissociation. Proc. IEEE **51**(1), 173–182 (1963)

Vitae

Mickey McDonald is a postdoctoral scholar in Cheng Chin's ultracold atoms group at the University of Chicago. After obtaining his bachelor's in physics at Cornell University in 2010, he pursued a PhD in atomic, molecular, and optical physics at Columbia University under the direction of Tanya Zelevinsky. He graduated in August 2016 as a Frances and Charles Townes Fellow. Recently, he won the 2017 Deborah Jin Award for Outstanding Doctoral Thesis Research in Atomic, Molecular, or Optical Physics.

CPSIA information can be obtained
at www.ICGtesting.com
Printed in the USA
LVHW02*1448040318
568593LV00001B/135/P